Gasoline Prices: Policies, Practices and Prospects

Gasoline Prices: Policies, Practices and Prospects

Barbara V. Urban
Editor

Novinka Books
New York

For permission to use material from this book please contact us:
Telephone 631-231-7269; Fax 631-231-8175
Web Site: http://www.novapublishers.com

NOTICE TO THE READER

Library of Congress Cataloging-in-Publication Data:
Available Upon Request

ISBN: 1-59454-651-7

Published by Nova Science Publishers, Inc. ✢ New York

CONTENTS

PREFACE

As major energy legislation moved to conference, the high price of gasoline remained a major consideration. The legislative proposals of past Congresses have contained numerous provisions that would affect gasoline supply and demand. This is true also of the Energy Policy Act of 2005, H.R. 6, both the version passed by the House April 21, and the Senate bill, passed June 28. A large number of factors combined to put pressure on gasoline prices, including increased world demand for crude oil and U.S. refinery capacity inadequate to supply gasoline to a recovering national economy. The war and continued violence in Iraq added uncertainty and a threat of supply disruption that added pressure particularly to the commodity futures markets. Numerous provisions in legislative proposals in the 108th Congress addressed perceived problems in the oil and gasoline markets. A comprehensive energy policy bill was reported out of conference and approved by the House, but several issues kept the bill from passing the Senate. Among the most controversial were provisions regarding the use of ethanol and the additive methyl tertiary butyl ether (MTBE) in motor fuel, proposals to open up part of the Arctic National Wildlife Refuge (ANWR) to oil and gas development, measures concerning corporate average fuel economy (CAFE) standards, and proposals to aid construction of new refineries and to harmonize state "boutique fuels" standards. In the 109th Congress, the House passed a comprehensive bill, H.R. 6, with many of the same provisions of the bill considered in the previous Congress. As before, MTBE and ANWR, included in the House-passed bill, remain controversial. The House bill added another controversial provision, giving the Federal Energy Regulatory Commission (FERC) overriding authority over state entities in licensing terminals to receive and process liquefied natural gas. In the Senate version of H.R. 6, the MTBE safe harbor provision has been omitted. The Senate bill contains a provision, not in the House-passed version, directing the President to take measures to reduce total demand for petroleum by one million barrels per day (mbd) by 2015. An amendment by Senator Cantwell, which would have set the goal of reducing petroleum imports by 40% by 2025, was defeated on the floor by a vote of 47-53. The gasoline price surge heightened discussion of energy policy, but the urgency of previous energy crises has been lacking. In part this may be due to the fact that there has been no physical shortage of gasoline, and no lines at the pump. In addition, the expectation of former crises, that prices were destined to grow ever higher, has not been prevalent. However, the persistence of high gasoline and oil prices into a second summer has raised alarms over the economic consequences of the situation.

Chapter 1

GASOLINE PRICES: POLICIES AND PROPOSALS[1]

Carl E. Behrens and Carol Glover

SUMMARY

As major energy legislation moved to conference, the high price of gasoline remained a major consideration. The legislative proposals of past Congresses have contained numerous provisions that would affect gasoline supply and demand. This is true also of the Energy Policy Act of 2005, H.R. 6, both the version passed by the House April 21, and the Senate bill, passed June 28.

A large number of factors combined to put pressure on gasoline prices, including increased world demand for crude oil and U.S. refinery capacity inadequate to supply gasoline to a recovering national economy. The war and continued violence in Iraq added uncertainty and a threat of supply disruption that added pressure particularly to the commodity futures markets.

Numerous provisions in legislative proposals in the 108th Congress addressed perceived problems in the oil and gasoline markets. A comprehensive energy policy bill was reported out of conference and approved by the House, but several issues kept the bill from passing the Senate. Among the most controversial were provisions regarding the use of ethanol and the additive methyl tertiary butyl ether (MTBE) in motor fuel, proposals to open up part of the Arctic National Wildlife Refuge (ANWR) to oil and gas development, measures concerning corporate average fuel economy (CAFE) standards, and proposals to aid construction of new refineries and to harmonize state "boutique fuels" standards.

---

[1] Excerpted from CRS Report IB10134 dated July 1, 2005.

In the 109th Congress, the House passed a comprehensive bill, H.R. 6, with many of the same provisions of the bill considered in the previous Congress. As before, MTBE and ANWR, included in the House-passed bill, remain controversial. The House bill added another controversial provision, giving the Federal Energy Regulatory Commission (FERC) overriding authority over state entities in licensing terminals to receive and process liquefied natural gas. In the Senate version of H.R. 6, the MTBE safe harbor provision has been omitted. The Senate bill contains a provision, not in the House-passed version, directing the President to take measures to reduce total demand for petroleum by one million barrels per day (mbd) by 2015. An amendment by Senator Cantwell, which would have set the goal of reducing petroleum imports by 40% by 2025, was defeated on the floor by a vote of 47-53.

The gasoline price surge heightened discussion of energy policy, but the urgency of previous energy crises has been lacking. In part this may be due to the fact that there has been no physical shortage of gasoline, and no lines at the pump. In addition, the expectation of former crises, that prices were destined to grow ever higher, has not been prevalent. However, the persistence of high gasoline and oil prices into a second summer has raised alarms over the economic consequences of the situation.

MOST RECENT DEVELOPMENTS

As the Senate finished consideration of H.R. 6, the Energy Policy Act of 2005, on June 23, crude oil prices touched $60 per barrel for the first time on the New York Mercantile Exchange, reflecting projections of continued strong world demand and uncertainty about production policies and capabilities of the Organization of Petroleum Exporting Countries (OPEC). The Senate passed H.R. 6 June 28 by a vote of 85-12. The House passed its version of the bill on April 21.

BACKGROUND AND ANALYSIS

The run-up of gasoline prices that began in spring 2004 (see **Figure 1**) climaxed a period of almost five years during which gasoline prices demonstrated a great deal of regional volatility but less of an increase at the national level. In 2004 a large number of factors combined to exert pressure on gasoline prices in all parts of the country. Some of these factors have affected the price of crude oil, and others the cost of producing and marketing gasoline.

Past energy crises have demonstrated that oil is traded in a world market, in which events in remote areas affect the price of crude for almost everyone. In the 12-18 months leading up to the crisis, these events included:

- Decisions by the Organization of Petroleum Exporting Countries (OPEC) cartel, after having reduced production quotas in 2002, to raise them only slowly and reluctantly;
- Unexpected demand growth in China;
- Disruptions in oil production in major exporters, including Venezuela, Iraq and Nigeria;
- Decline in the value of the U.S. dollar, the currency in which oil is traded in the world market, compared to other major currencies, particularly the Euro.
- Uncertainty and fear of major disruptions in Iraq and Saudi Arabia, in the context of the war in Iraq and the threat of terrorism.

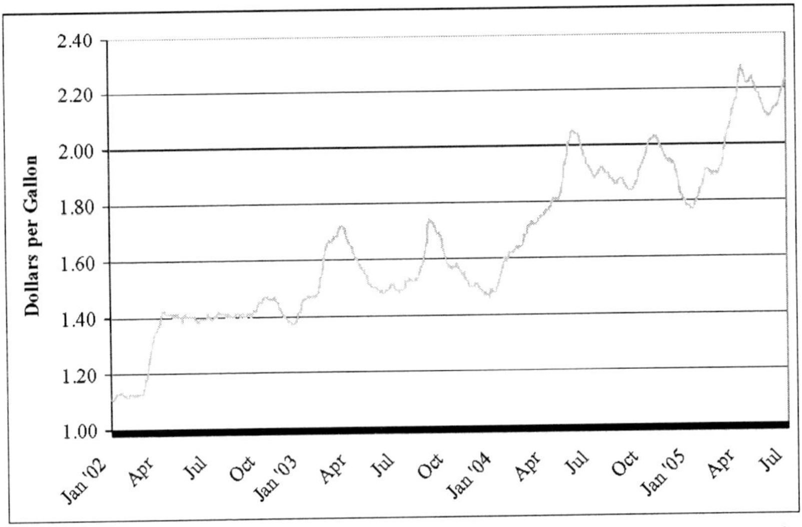

Figure 1. Average Daily Nationwide Price of Unleaded Gasoline, January 2002 - July 2005.

Note: Prices include federal, state and local taxes.

Source: *Daily Fuel Gauge Report*, American Automobile Association, [http://www.fuelgaugereport.com], compiled by CRS.

As often happens when commodity prices are volatile, speculation in futures contracts accentuated the upward price pressure and appeared to continue high prices longer than would be expected as market fundamentals push toward lower prices. Secretary of Energy Spencer Abraham, criticizing speculation in oil markets, asserted in July that the price of oil was $10 per barrel too high because of the possibility of disruptions in supply.[1] Nevertheless, the price went even higher in August and again in October, and continued high into the new year.

Just as a number of factors led to increased crude prices, a combination of features in the U.S. refinery industry contributed to an increase in gasoline prices.

- U.S. demand for gasoline has increased as economic growth has resumed.

- Domestic refining capacity has declined, both in number of refineries —from 324 in 1981 to 153 in 2002 — and in total capacity — from 18.62 million barrels per day (mbd) in 1981 to 16.78 mbd in 2002.

- The structure of the refining industry has changed. In 1981 most refining capacity was owned and operated by integrated oil companies that supplied 2 their own crude oil, refined it, distributed it, and marketed the products. Refining was only one part of the company's profit-making operation, and requently was not an important profit maker. Now the refining industry is haracterized more by independently owned, nonintegrated firms. When refineries are the sole source of revenue to the owners, it becomes more important that the operation be profitable, leading to pressure to raise prices.

- The refining industry has been operating with lower inventories of both crude oil and gasoline, as a means of cutting costs. The side effect has been reduced ability to meet unanticipated demand, leading to greater price pressure.

- Gasoline markets are fragmented regionally because air quality requirements have led to numerous different formulations to meet varying standards. In meeting demand for these regional formulations, called "boutique fuels," refiners lose flexibility to meet local variations in demand elsewhere, leading to increased price pressure.

- With domestic refining capacity constraints, a greater proportion of gasoline demand is being met with imported products. Foreign refiners typically manufacture products designed to sell in the international market, not the special product "boutique fuels" demanded by a significant share of the U.S. market.

- Refiners have had increased costs in the past year to comply with new requirements to limit sulfur content and to switch from the oxygenate additive MTBE to ethanol.

These various factors pushed the nationwide average price of gasoline over $2 per gallon in May 2004. By mid-June, Energy Information Administrator Guy Caruso was able to note a slight decline in prices, and tell a Senate Energy Committee hearing that, "absent major disruptions, oil and gasoline markets may be turning a corner."[2] However, persistent high crude prices pushed gasoline prices over $2 again in October, and yet again in March 2005. By April 2005, Caruso was suggesting that increasing world demand for oil might keep the price of crude above $50 per barrel through 2006.[3]

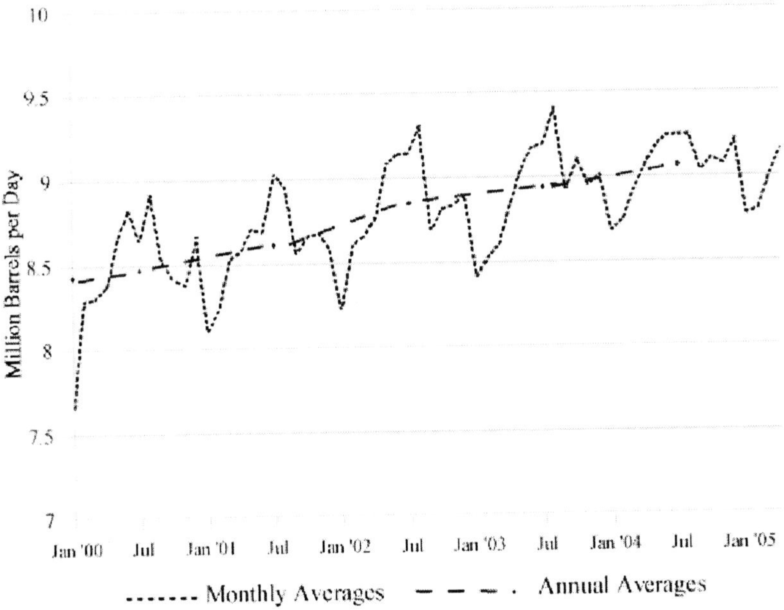

Figure 2. Consumption of Motor Gasoline, 2000-April 2005

Source: EIA, *Monthly Energy Review*, January 2003, January 2004 and May 2005, Table 3.4.

The price surge intensified discussion of energy policy and led to further calls for passage of energy legislation. However, the urgency of previous energy crises has been lacking. Throughout the period, U.S. gasoline consumption has continued to rise, although the usual summer peak in consumption appears to have been somewhat blunted in 2004, as shown in **Figure 2**. In part this may be because, although the price of gasoline in nominal terms set a record, in real terms it was less than in the Iranian crisis years of the early 1980s (see **Figure 3**). Further, unlike the earlier crises, there was no physical shortage of gasoline, and no lines at the pump. In addition, as **Figure 4** indicates, the proportion of consumer expenditure on oil and gasoline had declined from the high levels of the 1970s and early 1980s. Perhaps most important, the common view during the earlier crises was that oil prices not only were high, but were destined to become ever higher in the coming years. This view is no longer prevalent, and the general expectation has been that the run-up of prices in 2004 is a temporary phenomenon, although lasting longer than expected.

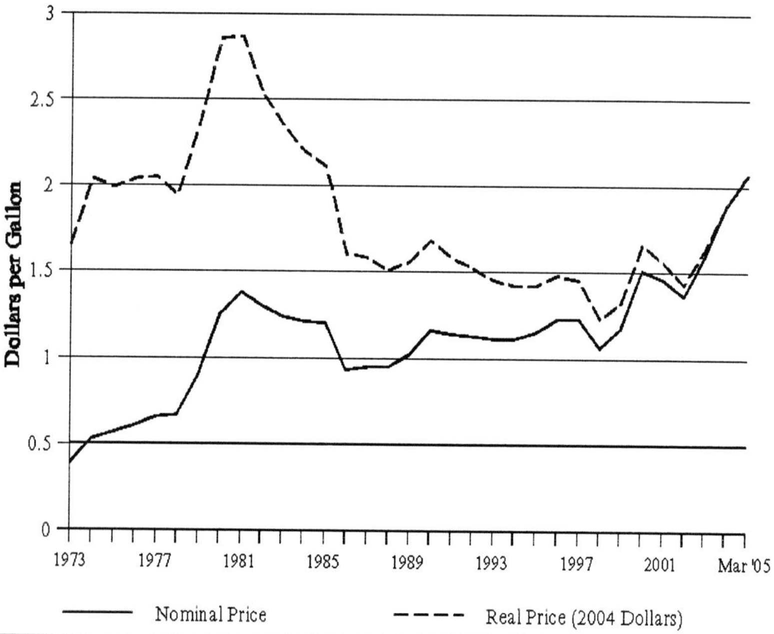

Figure 3. Nominal and Real Price of Gasoline, 1973-2004 and March 2005

Source: EIA, *Monthly Energy Review*, May 2005, Tables 1.6 and 9.4, calculated by CRS.

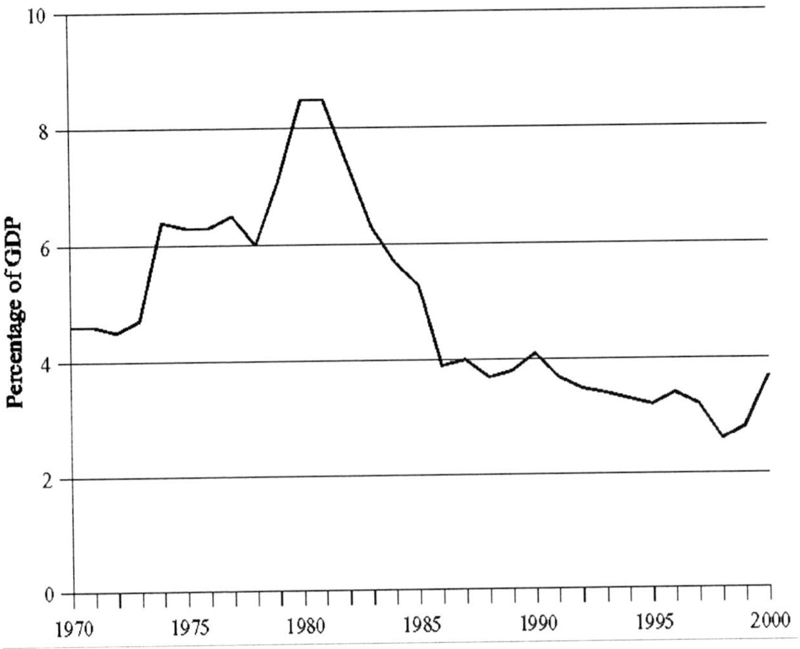

Figure 4. Consumer Spending on Oil as % of GDP, 1970-2000

Source: EIA, Annual Energy Review 2003, Table 3.4.

As shown in **Figure 5**, gasoline prices historically have increased less than the general rate of inflation, as measured by the Consumer Price Index (CPI). After the surge in 1973, and again after the 1979-1980 run-up, gasoline prices grew very slowly and even declined, dropping sharply in 1986. A sudden increase in 2000 was similarly followed by slow or declining prices. During the current run-up, for which data are not yet available, gasoline price increases have far outpaced the general CPI increase.

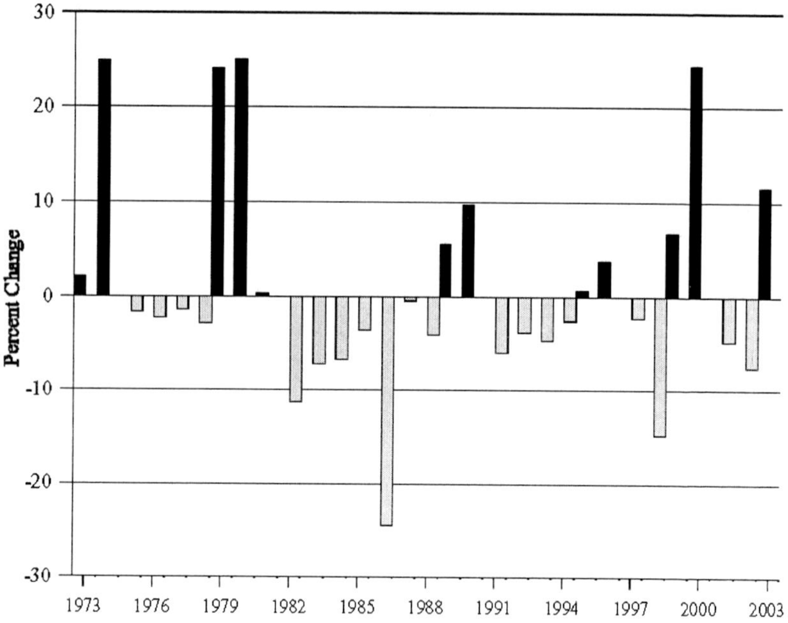

Figure 5. Percent Change in Gasoline Prices Compared to the Consumer Price Index, 1973-2003

Source: EIA, Annual Energy Review 2003, Table 5.24. CPI from the Bureau of Labor Statistics. Calculated by CRS.

Policy Options

The several energy crises of the past led to major legislative action, twice in the 1970s and once following the 1991 Gulf War. The current legislative situation differs from the previous actions because the Congress had been considering major energy legislation for three years before the situation became a nationwide concern. The various versions of a comprehensive energy bill, which failed to pass in the 108th Congress, all contained measures addressing some of the problems putting pressure on gasoline prices. In addition, a number of stand-alone legislative proposals to deal with some of the specific problems noted above were introduced in the 108th Congress. Most of these measures were included in the 109th Congress's energy bill, the Energy Policy Act of 2005, H.R. 6, which passed the House April 21, 2005, and passed the Senate, with significant differences, on June 28.

As in previous legislative energy debates, a major policy divide exists between those who view the gasoline-fueled automobile as a temporary necessity to be tolerated only until a substitute fuel or alternative means of transportation can be developed, and those who expect oil to be the same dominant transportation fuel in the indefinite future that it is at present. Compromise agreements have been reached via a combination of measures that enhance the development of alternatives or restrain the growth in demand for oil, on the one hand, and those that increase production or reduce the cost of supplying that demand, on the other. However, individual measures often carry with them complicating features that make passage more difficult. In addition, major legislation often becomes the vehicle for measures that typically would not find enough support to pass as individual bills, or which may be added to gain support for the whole measure. In the legislative climate of the 108th Congress, balancing the various interests involved proved too difficult a task, despite the influence of a nationwide energy crisis in an election year. Under the stimulation of continued high oil and gasoline prices, the 109th Congress is pursuing the goal again, with supporters of the legislation expressing more optimism that conflicts can be resolved.

Although major legislation did not pass in the 108th Congress, a number of the energy tax measures were included in two tax bills that became law, the Working Families Tax Relief Act (H.R. 1308, P.L. 108-311) and the American Jobs Creation Act (H.R. 4520, P.L. 108-357). However, in the 109th Congress H.R. 6 as passed by the House contains numerous tax measures, many of them affecting the oil industry. In the Senate, a package of tax measures reported by the Senate Finance Committee was attached to H.R. 6 on the floor before the bill was passed June 28.

Major Oil-Related Issues for the 109th Congress

A number of issues have been major barriers to passage of omnibus energy legislation, and remain to be resolved in the 109th Congress. Some of these issues do not involve oil; provisions to continue the restructuring of the electric power industry, for instance, have been and continue to be controversial. Among oil-related issues, proposals to open part of the Arctic National Wildlife Refuge (ANWR) to oil and gas development, and measures concerning Corporate Average Fuel Economy (CAFE) standards have stimulated major debate. A primary stumbling block has been the issue involving ethanol as an automobile fuel, and the related problems involving the gasoline fuel additive MTBE.

Ethanol and MTBE

The roots of the controversy lie in the Clean Air Act Amendments of 1990, which mandated that "reformulated" gasoline required in some localities to improve air quality contain 2% oxygen. This requirement could be met by adding ethanol to gasoline, but it could also be achieved by adding a substance called methyl tertiary butyl ether (MTBE), which had been produced in small quantities for many years as an octane enhancer. Because MTBE was cheaper than ethanol and was easier to mix and transport than ethanol, many refiners began using it to meet the new standards.

However, as its use spread, it became apparent that MTBE tended to escape easily from pipes and storage tanks, and contaminate water supplies, imparting a taste and odor that was unpalatable even in small quantities. This development led to moves to restrict and prohibit the use of MTBE. It also led a number of communities to sue refiners for the cost of decontaminating their water supply. At the same time, evidence began to accumulate that oxygenating gasoline was not necessary to achieve the air quality benefits of reformulated gasoline.

The omnibus energy bills in the 108th Congress addressed this changing situation by repealing the oxygenation requirement in the Clean Air Act, but adding a new mandate that gasoline have an increasing amount of renewable fuel, presumably ethanol. Consumption of ethanol in gasoline in 2002 was 2.1 billion gallons. Under the Renewable Fuel Standard, the amount required to be consumed would be 3.1 billion gallons in 2005 and 5.0 billion gallons by 2012. This would still be a small proportion of the total amount of gasoline consumed, which was close to 150 billion gallons in 2004, but was expected to stimulate the ethanol industry and the agricultural sector that supplies it. It was opposed by oil industry interests, who complained of the mandated increase in consumption of ethanol, which receives a substantial tax credit. Some suggested that it would raise prices locally, despite the subsidy.

The most controversial measure in the bills was a so-called "safe harbor" provision from product liability lawsuits for producers of MTBE and renewable fuels. It was a major factor in the failure of comprehensive energy legislation in the Senate in the 108th Congress.

In the 109th Congress, H.R. 6, as reported by the House Committee on Energy and Commerce on April 13, 2005, retained the safe harbor provision; an amendment to remove it was defeated in committee. When the bill reached the House floor, Representative Capps brought up an amendment to strike the safe harbor language. After extended debate specifically on the MTBE provision, the Capps amendment was narrowly defeated by a vote of 213-219. The Senate version of H.R. 6 does not contain the safe-harbor provision.

ANWR

Oil and gas exploration and development of part of the Arctic National Wildlife Refuge have been controversial for many years. This was part of the early proposals for legislation that eventually became the Energy Policy Act of 1992, but was dropped in the face of strong opposition in both houses.

Support for action grew gradually through the decade, along with technological developments that advocates claimed would reduce the environmental impact of development, and the House included a development measure in its version of an omnibus energy bill in August 2001. A similar measure was part of the House-passed legislation in the 108th Congress. Opposition in the Senate kept the measure from the floor, however, and it was dropped in conference.

In the 109th Congress, Senate supporters of ANWR development have moved the issue to the budget process, where it can be approved by a simple majority vote. On March 9, 2005, the Senate Budget Committee issued a FY2006 budget resolution that assumes $2.5 billion of revenue over five years from leases in ANWR. On March 16 the Senate rejected an amendment by Senator Cantwell to strike the ANWR provisions, by a vote of 49-51. The next day the Senate passed the budget resolution (S.Con.Res. 18).

The House FY2006 budget resolution (H.Con.Res. 95, passed March 17) does not contain the ANWR provision. However, the Energy Policy Act of 2005, H.R. 6, does authorize exploration and development of ANWR, and an amendment to strike the provision during floor debate was defeated, 200-231.

CAFÉ

Fuel economy standards also have a long history of controversy, going back to their establishment in the 1970s. Proposals to mandate new standards were also considered, but dropped, early in the development of the 1992 Energy Policy Act. In the mid-1990s the National Highway Traffic Safety Administration (NHTSA) was considering a rulemaking that would result in increased standards for light duty trucks (including sport utility vehicles), but for several years the Congress included in its annual appropriation for NHTSA a measure prohibiting NHTSA from analyzing or undertaking such a ruling. That prohibition was dropped in the FY2004 NHTSA appropriations, and a final rule issued by NHTSA in April 2003 requires a boost in light truck fuel economy to 22.2 miles per gallon by Model Year 2007.

Omnibus energy legislation proposed before NHTSA acted would have mandated specific increases in light truck fuel economy, but in the 109th Congress H.R. 6 would merely amend slightly the criteria NHTSA must follow in its rulemaking and authorize appropriations of $2 million annually through FY2008 for that purpose. During floor debate on H.R. 6, an amendment to increase fuel economy standards to 33 miles per gallon over 10 years was defeated by a vote of 177-254.

A more general amendment to the House bill, requiring the Administration to take "voluntary, regulatory, and other actions" to reduce oil demand in the United States by 1 million barrels per day from projected levels by 2013 was defeated 166-262. However, the measure was included in the bill reported out by the Senate Energy Committee. An amendment by Senator Cantwell, which would have set the goal of reducing petroleum imports by 40% by 2025, was defeated on the floor by a vote of 47-53.

The NOPEC bill

Frustration with high gasoline prices was reflected in adoption, as an amendment to H.R. 6, of Senator DeWine's No Oil Producing or Exporting Cartel (NOPEC) Act of 2005, S. 555. The bill would declare limiting production or setting prices of oil by a foreign state as illegal in U.S. courts, deny sovereign immunity to such states, and authorize the Attorney General and the Federal Trade Commission to bring action against them. The amendment was adopted by voice vote June 21. Senator Domenici noted after the vote that the House version of H.R. 6 does not contain a NOPEC provision and warned that it might not survive conference.

Other Oil and Gasoline Measures

Other provisions in H.R. 6 related to petroleum would authorize:

- The federal government to continue to receive physical quantities of oil and gas as royalty-in-kind payments instead of cash payments for royalties on leased federal property.
- Royalties for certain types of leases in the Gulf of Mexico, and others such as marginal wells to be lowered or terminated.
- Regulatory requirements to be eased for some oil and gas activities such as hydraulic fracturing and construction of exploration and production facilities.
- The system of leasing and permitting access to federal lands for oil and gas development to be amended.
- The amendment of statutes concerning alternative-fueled vehicles.
- Steps to discourage the proliferation of state "boutique fuels" requirements. Additionally, a study of "harmonization" of current fuel controls would be mandated.

In June 2004 in the last Congress, the House passed H.R. 4517, the U.S. Refinery Revitalization Act, which would have eased regulatory requirements for construction of new refineries in areas of high unemployment. In the 109th Congress, the provisions of the Refinery

Revitalization Act were included in H.R. 6 as passed by the House April 21, 2005.

A new provision in H.R. 6 gives the Federal Energy Regulatory Commission (FERC) authority over siting and licensing of facilities to import and process liquefied natural gas (LNG). The statutory language would largely confirm FERC's assumption of a lead role in siting LNG facilities, which has been challenged in court by several states. An amendment to strike the LNG provision was defeated on the House floor by a vote of 194-237.

REFERENCES

[1] "Abraham criticizes high cost of oil," *Washington Times*, July 7, 2004, p. A1.
[2] Guy Caruso, statement before the Senate Committee on Energy and Natural Resources, June 15, 2004.
[3] "EIA: Oil to remain above $50 through '06." *Oil Daily*, April 8, 2005, p. 1. Report of a speech by Guy Caruso at the National Press Club in Washington, D.C., April 7, 2005.

Chapter 2

MOTOR FUELS: UNDERSTANDING THE FACTORS THAT INFLUENCE THE RETAIL PRICE OF GASOLINE*

General Accounting Office

ABBREVIATIONS

API: American Petroleum Institute:

BLS: Bureau of Labor Statistics:

EIA: Energy Information Administration:

FTC: Federal Trade Commission:

OPEC: Organization of Petroleum Exporting Countries:

SUV: sport utility vehicle:

SECTION 1: BACKGROUND:

Q. How much of a barrel of oil is made into gasoline?

A. Gasoline is one of several petroleum products made by refining crude oil. As Figure 1 shows, on average, gasoline constitutes almost half of the

---

* Excerpted from GAO-05-525SP dated May 2005

total petroleum products derived from a barrel of crude oil, although the amount of gasoline that can be produced from a barrel of crude oil varies among individual refineries.

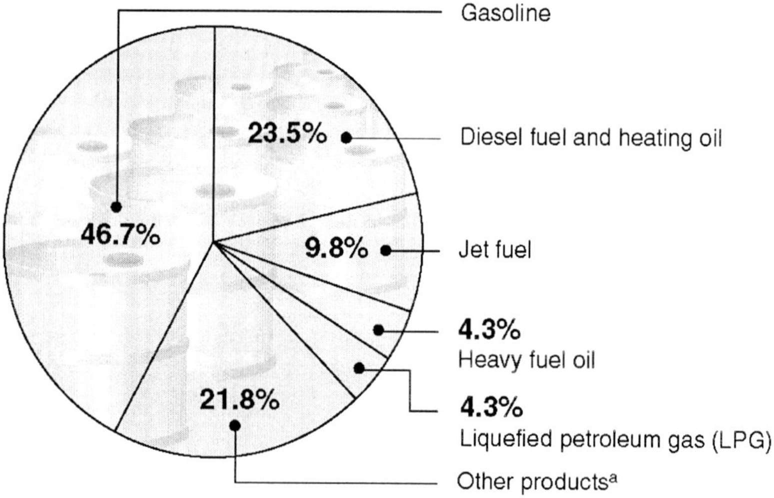

Figure 1: Products Made from a Barrel of Crude Oil (Average for 2000 to 2004).

Source: GAO analysis of EIA data.

Notes: A barrel containing 42 U.S. gallons of crude oil yields slightly more than 44 gallons of petroleum products. This "process gain" in volume is due to a reduction in density during the refining process. As a result, percentages total more than 100 percent.

[a] Other products include petroleum coke, still gas, asphalt and road oil, petrochemical feedstock, lubricants, etc.

Overall, the amount of gasoline a refiner can produce from a barrel of crude oil depends, in part, on the configuration (sophistication) of the refinery. More sophisticated refineries can squeeze more gasoline out of a given barrel of crude oil than less sophisticated ones. Refineries in the United States are among the most sophisticated in the world. The amount of gasoline produced also depends on the quality of crude oil: light crude oil, which is generally low in sulfur, can yield more gasoline than heavy crude oil, which tends to be high in sulfur.[Footnote 3]

Q. How does gasoline get to the consumer?

A. As Figure 2 shows, gasoline reaches the consumer through a complex system that begins with extracting crude oil and transporting it to refineries, mostly via pipelines, marine tankers, and barges. [Footnote 4] At the refineries, crude oil is processed into gasoline and other petroleum products. The gasoline is then transported--again, usually via pipelines, marine tankers, and barges--to storage terminals for wholesale distribution. From there, it is shipped by truck to retail stations, where consumers pull up to fill their tanks.

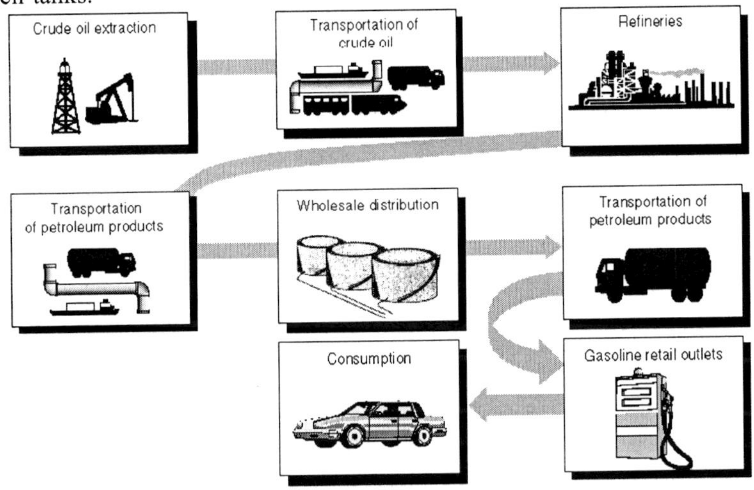

Figure 2: Gasoline Production and Distribution System:
Source: GAO.

Q. How have gasoline prices changed over time?

A. As shown in Figure 3, the average U.S. retail price for gasoline has nearly quadrupled since 1974.

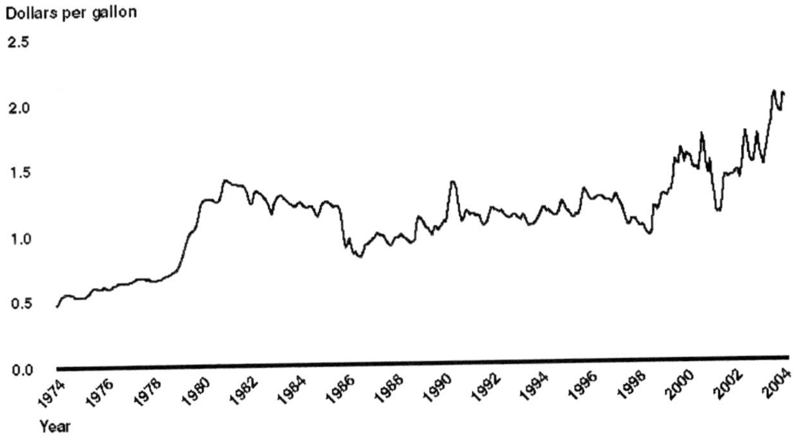

Figure 3: U.S. Retail Price of Gasoline (1974-2004):
Source: GAO analysis of Bureau of Labor Statistics (BLS) data.

Note: Price data are monthly U.S. city averages for regular gasoline. Prices in 1974 and 1975 are for leaded gasoline, and prices for 1976 onward are for unleaded gasoline.

However, as shown in Figure 4, the average U.S. retail gasoline price adjusted for inflation was actually substantially lower in 2004 than it was during its peak in 1981. Even so, the average U.S. retail gasoline price has experienced frequent ups and downs, or volatility, and was significantly higher in 2004 than it was as recently as 2002. In fact, the average U.S. gasoline price in 2004, adjusted for inflation, was the highest since 1985.

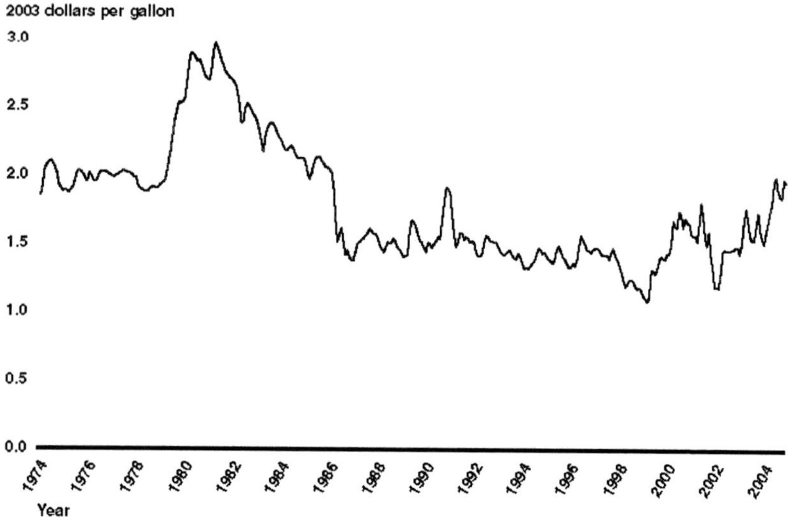

Figure 4: U.S. Retail Price of Gasoline Adjusted for Inflation (1974- 2004):

Source: GAO analysis of BLS data.

Note: Price data are monthly U.S. city averages for regular gasoline. Prices in 1974 and 1975 are for leaded gasoline, and prices for 1976 onward are for unleaded gasoline. Gasoline prices were adjusted to 2003 dollars using the Bureau of Labor Statistics' Consumer Price Index for All Urban Consumers: All Items.

No one factor can entirely explain the variations in gasoline prices. As will be discussed in the following sections, the dynamics of gasoline prices are a function of many complex factors that are sometimes interconnected. Before explaining each key factor and the underlying causes of its behavior, we first examine the key elements that make up the price of a gallon of gasoline.

SECTION 2: ELEMENTS OF GASOLINE PRICE:

Q. What do consumers pay for in a gallon of gasoline?

A. According to EIA, four elements determine the price consumers pay for a gallon of gasoline at a given time: (1) the price of crude oil, (2) taxes, (3) refining costs and profits, and (4) distribution and marketing costs and profits. Figure 5 shows the average percentage of each of these elements in the U.S. price of a gallon of gasoline in 2004.

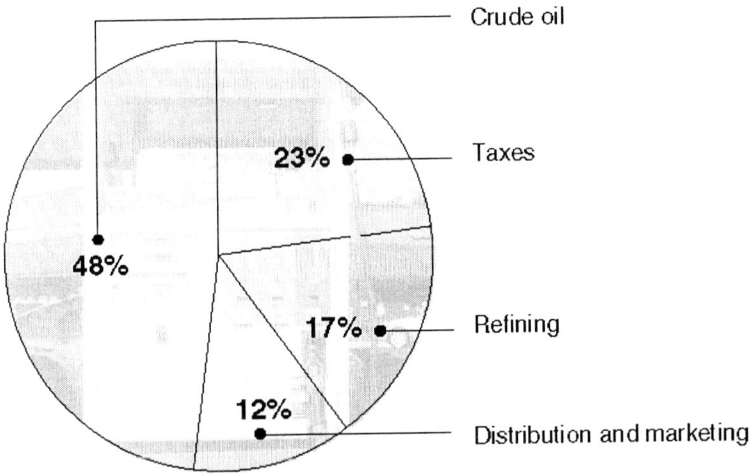

Figure 5: Elements of Gasoline Price (2004 Average):
Sources: GAO analysis of EIA data; GAO (photo).

Crude Oil:

According to EIA's data, the price that refiners paid for crude oil in 2004 was the largest component of the price of regular gasoline. On average, crude oil price accounted for slightly less than half of the average price of a gallon of regular gasoline. However, the proportion of the price of gasoline that crude oil constitutes is not constant over time, nor is the share of the other elements as shown in Figure 6.

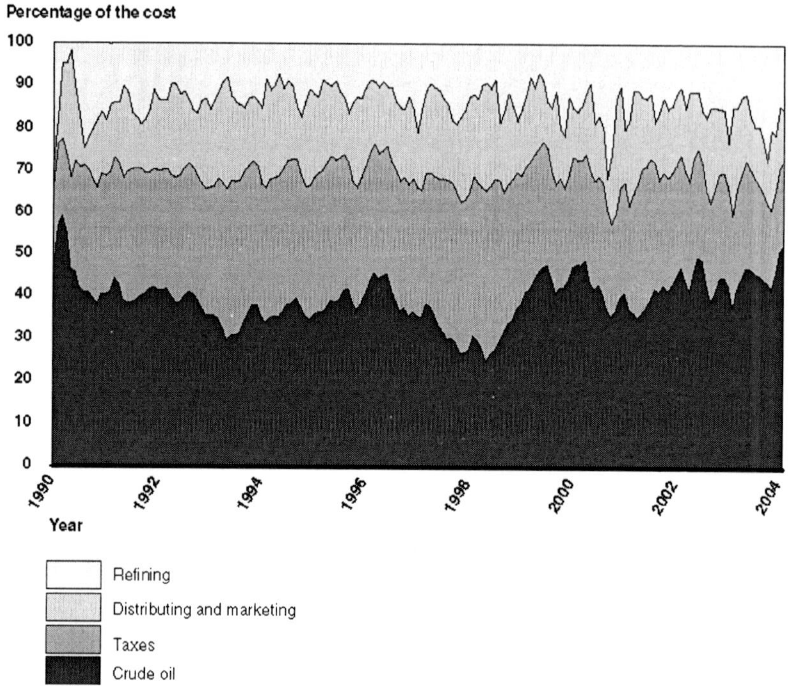

Figure 6: Elements of Gasoline Price (1990-2004):

Source: GAO analysis of EIA data.

Taxes:

On average, taxes accounted for 23 percent of what consumers paid for a gallon of gasoline in 2004, according to EIA's data. This percentage includes estimated federal and average state taxes totaling 44 cents per gallon (see Fig. 7).[Footnote 5] Federal taxes accounted for 18.4 cents of this total, while state taxes averaged 25.6 cents per gallon, although taxes vary among states. A chart showing total gasoline taxes for each of the 50 states and the District of Columbia is presented in section 4 of this primer.

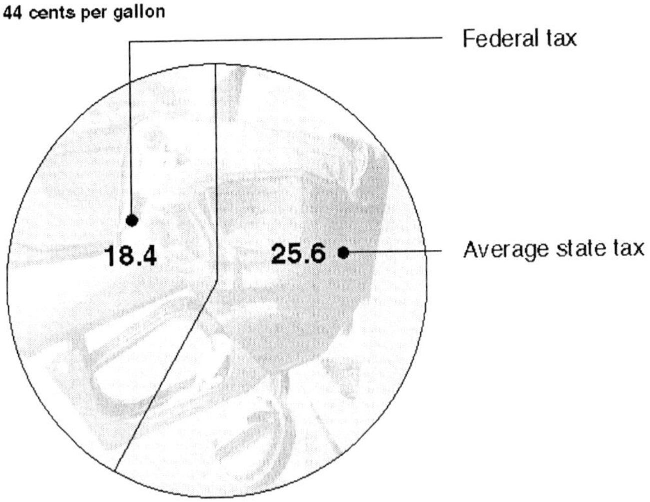

Figure 7: Estimated Federal and Average State Gasoline Taxes Per Gallon (2004):
Sources: GAO analysis of API data; GAO (photo).

Refining:

On average, refining costs and profits made up about 17 percent of the pump price for regular gasoline in 2004, based on EIA data. This element of the pump price includes the labor, materials, energy, and other costs of refining crude oil through a number of physical and chemical transformations to create the products in Figure 1. It also includes the profits to refinery owners.

Distribution and Marketing:

According to EIA data, the costs and profits associated with distribution and marketing of gasoline constituted about 12 percent of the average price of a gallon of regular gasoline in 2004. This price element includes the costs of distributing and marketing gasoline from refiners to wholesale marketers and/or retail marketers and the profits made by the wholesale and retail marketers.

While we have provided a snapshot of the key components of the retail price of gasoline at any given point in time, the snapshot approach does not explain why the price that consumers pay changes over time and varies across geographic areas, nor does it explain the influences behind this behavior. Sections 3 and 4 provide more insight into the dynamic behavior of retail gasoline prices by examining the factors that cause these changes in their broader, and often global, contexts.

SECTION 3: VARIATIONS IN GASOLINE PRICES OVER TIME:

Q. What are the key factors causing gasoline prices to change over time?

A. A number of key factors cause the retail price of gasoline to vary over time. First, and most importantly, gasoline price changes are caused by changes in the price of crude oil. As Figure 8 shows, the trend in gasoline prices has generally tracked the trend in crude oil prices, although the two do not necessarily rise and fall in lockstep. Second, gasoline prices vary because of changes in how much gasoline we demand relative to the supply. Third, federal and state requirements calling for special gasoline blends to improve air quality can make it more costly to produce gasoline, thus increasing its price. Fourth, seasonal factors affect demand and supply, and hence, the price of gasoline. Fifth, changes in taxes levied on gasoline have affected gasoline prices over the past decade, but because federal and state gasoline taxes remained relatively constant in recent years, they have not significantly contributed to recent volatility in gasoline prices. Sixth, changes within the petroleum industry, such as mergers, can also cause the price of gasoline to change over time. How these key factors are, in turn, affected by other influences is explained in the answers to subsequent questions in this section.

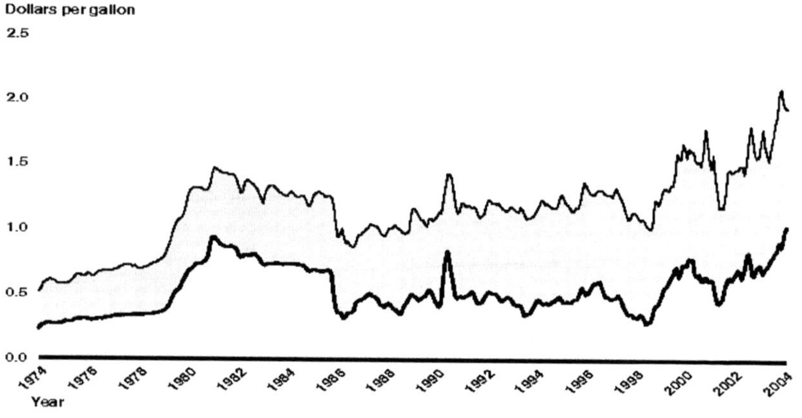

Figure 8: Crude Oil and Gasoline Prices (1974-2004):

Source: GAO analysis of EIA and BLS data.

Note: Crude oil price data are the monthly refiner acquisition cost of crude oil composite from EIA. Gasoline price data are the monthly U.S. city average price for regular gasoline from BLS.

Q. What determines the price of crude oil?

A. The price of oil is determined in the world market and depends mainly on the balance between world demand and supply. That balance is in turn influenced by a number of other factors, as discussed below.

World Oil Demand:

Because oil is purchased and consumed in many countries, including the United States, at a price determined in the world market, changes in world demand--not just U.S. demand--can influence the price that U.S. refiners pay for crude oil and, consequently, the price that U.S. consumers pay for gasoline. As Figure 9 shows, since 1983 world oil demand grew from about 59 million barrels per day to more than 82 million barrels per day in 2004, for an average annual increase of 1.9 percent.

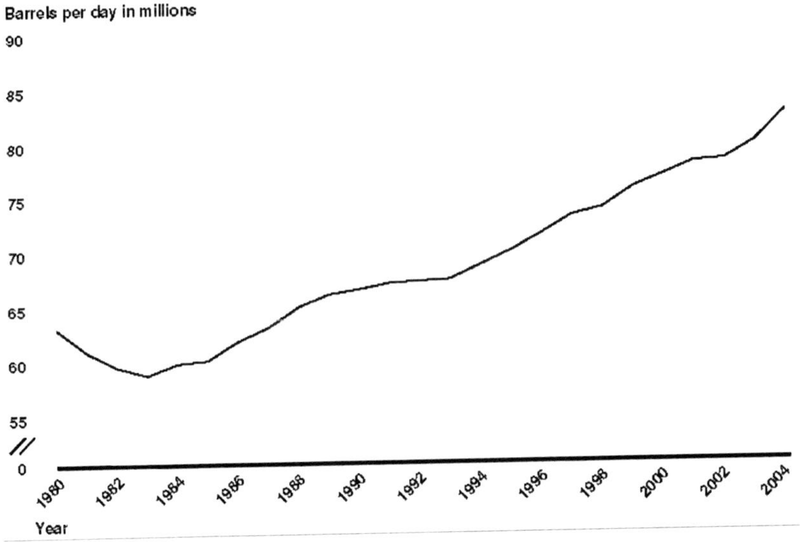

Figure 9: World Oil Demand (1980-2004):

Source: GAO analysis of EIA data.

The United States is the world's largest oil consumer, accounting for about 25 percent of the world's demand, although it constitutes only about 5 percent of the world's population. As Figure 10 shows, the United States consumed about 20.5 million barrels of petroleum per day in 2004, compared with about 15.2 million barrels per day in 1983, for an average annual increase of 1.65 percent.

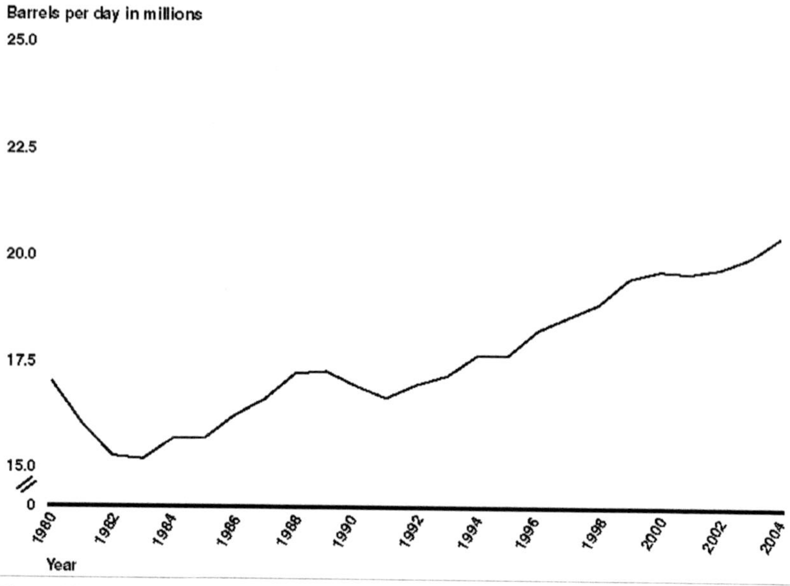

Figure 10: U.S. Oil Demand (1980-2004):

Source: GAO analysis of EIA data.

Recently, rapid growth in oil demand in Asia contributed to a rise in crude oil prices to more than $50 per barrel during 2004. Growth in demand was particularly strong in China where, according to a recent estimate by the National Commission on Energy Policy,[Footnote 6] demand grew by 18 percent during the first quarter of 2004, after rising 11 percent in 2003.

An important aspect of oil demand is the difficulty of substituting other sources of energy for some petroleum products--particularly gasoline. In the immediate term at least, gasoline has virtually no cost-effective substitutes, and, therefore, increases in its price cause only small decreases in consumption.

World Oil Supply:

Unlike oil demand, which encompasses demand from most of the world's nations, currently, relatively few countries supply most of the world's oil because those are the countries that have oil reserves and production. World oil producers can be grouped into two broad categories--members of the Organization of Petroleum Exporting Countries (OPEC) and non-OPEC countries. OPEC currently has 11 members, and many of its major producers are located in the Persian Gulf region.[Footnote 7] Key non-OPEC producers include Russia, the United States, China, Mexico, Norway, Canada, and the United Kingdom.

As Figure 11 shows, non-OPEC countries, including the United States, have typically contributed the majority of the world's oil supply since the 1970s. Over the same period, the U.S. share of total world supply has decreased from about 23 percent in 1970 to about 10 percent in 2004. The United States is importing increasingly larger shares of its oil use. In 2004, the United States imported about 13 million barrels of crude oil and petroleum products per day, representing about 60 percent of total U.S. consumption. About 56 percent of the imports were from non-OPEC countries, of which Canada, Mexico, the United Kingdom, and Angola were key suppliers. OPEC supplied the remaining 44 percent of U.S. imports, mostly from Saudi Arabia, Venezuela, Nigeria, and Iraq. The Persian Gulf region accounted for about 19 percent of total U.S. imports of crude oil and petroleum products.[Footnote 8]

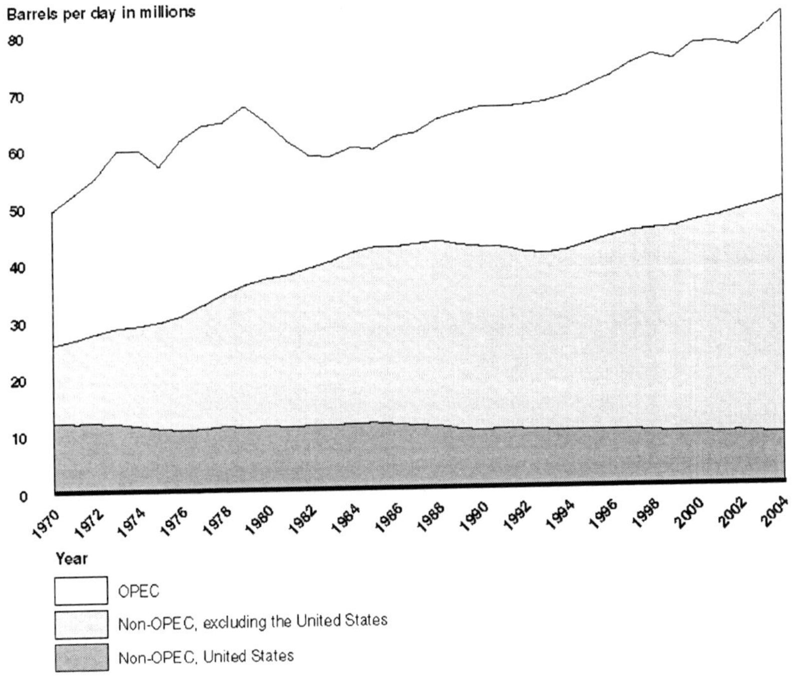

Figure 11: World Oil Supply (1970-2004):

Source: GAO analysis of EIA data.

While non-OPEC countries still contribute the majority of the world's oil supply, the share of world oil supplied by OPEC has increased since the mid-1980s and accounted for about 40 percent of world supply in 2004. Overall, OPEC supply makes up the balance of world oil demand over non-

OPEC supply. In its role as the world's demand-supply balancer, OPEC's supply decisions play a critical role in the determination of world oil price.

OPEC influences the world oil price mostly through its supply decisions. While the group had little or no influence on the world oil market during the first 10 years after it was formed in 1960, generally, it has been a dominant force in the world oil market since the 1970s. Its influence on the world price of crude oil has varied depending on the conditions of world supply and demand and its members' discipline in adhering to their production quotas. As a group, OPEC established a "target" price of $22 to $28 per barrel for its basket of crude oil in 2000.[Footnote 9] To achieve this target price, OPEC periodically meets and sets an aggregate production level, or quota, based on the organization's determination of the demand for its oil. (Fig. 12 shows an OPEC meeting.) OPEC then allocates voluntary production quotas among its members, primarily based on the size of each member's oil reserves. Whether the target price is achieved depends on the discipline exercised by member countries in producing oil, as well as on the actual demand for oil and non-OPEC countries' production levels. If OPEC's estimates of demand are correct and member countries adhere to their production quotas, the average world price will likely be close to the target price. On the other hand, if some or all members exceed their quotas and overproduce, the average world price may be substantially below the target price.

Figure 12: OPEC Meeting:

Source: OPEC.

If the world crude oil price exceeds the target price, OPEC may not always increase its production to bring the world price down to its target. For example, OPEC members decided in their December 2004 meeting to cut production as of January 1, 2005, to keep the world price from falling, even though the then-current price of around $36 per barrel for the OPEC basket of crude oil exceeded OPEC's target price of $22-28 per barrel.[Footnote 10] Two conditions give OPEC countries considerable flexibility to influence world oil prices:

* As a group, OPEC holds the world's largest, lowest-cost reserves of crude oil. OPEC countries account for over three-fourths of the world's estimated conventional reserves of about 1 trillion barrels (see Fig. 13). In contrast, the United States has reserves of about 22.7 billion, or slightly less than 2 percent of the world's total.

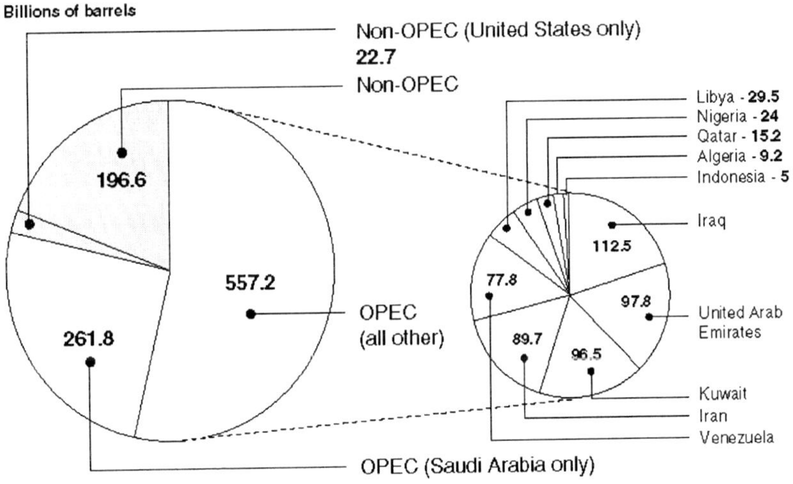

Figure 13: World Oil Reserves, OPEC and Non-OPEC (2003):

Source: GAO analysis of EIA data.

Note: This estimate is for conventional crude oil and does not include bitumen contained in oil sands in Canada.

* OPEC is the only entity that can increase crude oil production relatively quickly if there is supply shortage because it holds virtually all of the world's excess crude oil production capacity. In fact, according to EIA data, as of November 2004 Saudi Arabia had virtually all of the world's excess crude oil production capacity--up to 1 million barrels per day. (See Fig. 14.)

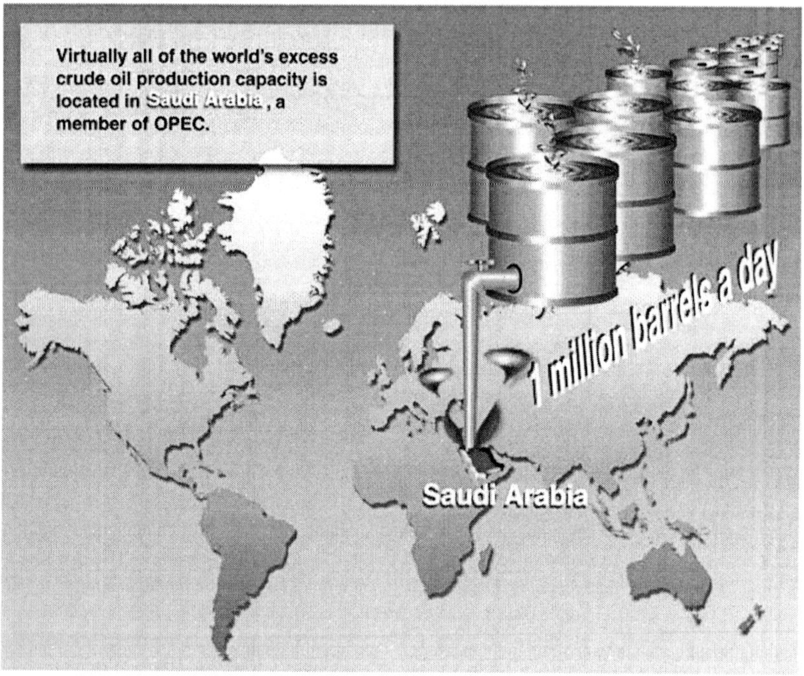

Figure 14: World Excess Crude Oil Production Capacity:

Source: GAO analysis of EIA data.

Apart from OPEC's supply decisions, OPEC countries sometimes experience events that affect their oil supplies and the world price of crude oil. For example, the current and past wars in Iraq and strikes in Venezuela and Nigeria disrupted oil supplies and temporarily raised the price of crude oil.

The futures market typically reflects the impact of world events on the world price of crude oil. For example, political instability and terrorist acts in countries that supply oil create uncertainties about future supplies. Faced with such uncertainties, oil traders at futures markets such as the New York Mercantile Exchange may bid up futures prices for oil in anticipation of an oil shortage and expected higher prices in the future.[Footnote 11] On the other hand, oil traders may also bid futures prices down in reaction, for example, to news about a new oil discovery that would increase world oil supply and potentially lower the world price. In other words, market participants' expectations of what may happen to oil supply and demand in the context of world events influence their price bids. (See Fig. 15.) Although crude oil buyers may not always pay the exact price observed in the futures markets, the futures markets play a critical role in providing a leading reference price for buyers' and sellers' negotiations in crude oil transactions.

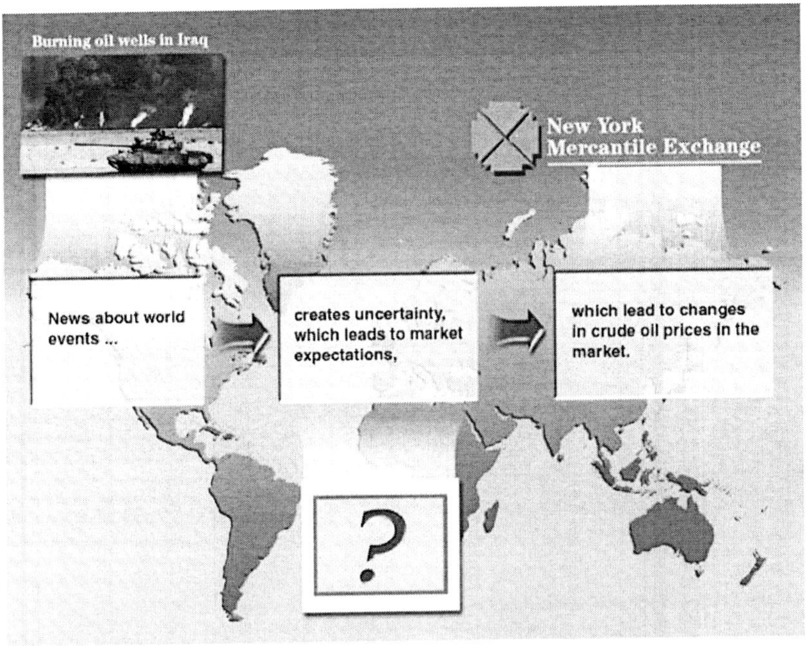

Figure 15: World Events, Market Expectations, and the Oil Futures Market:
Sources: GAO (analysis); U.S. Army (photo).

Q. How do changes in gasoline demand relative to supply affect the price?

A. As is the case for most goods and services, changes in the demand for gasoline relative to changes in supply affect the price that consumers pay. In other words, if the demand for gasoline increases faster than the ability to supply it, the price of gasoline will most likely increase.

In 2004, the United States consumed an average of about 380 million gallons (about 9 million barrels) of gasoline per day. This consumption is about 56 percent more than the 1970 average per day consumption of 243 million gallons (about 5.8 million barrels)--an average increase of about 1.6 percent per year for the last 35 years. U.S. gasoline consumption has increased particularly steeply since the 1990s. (See Fig. 16.)

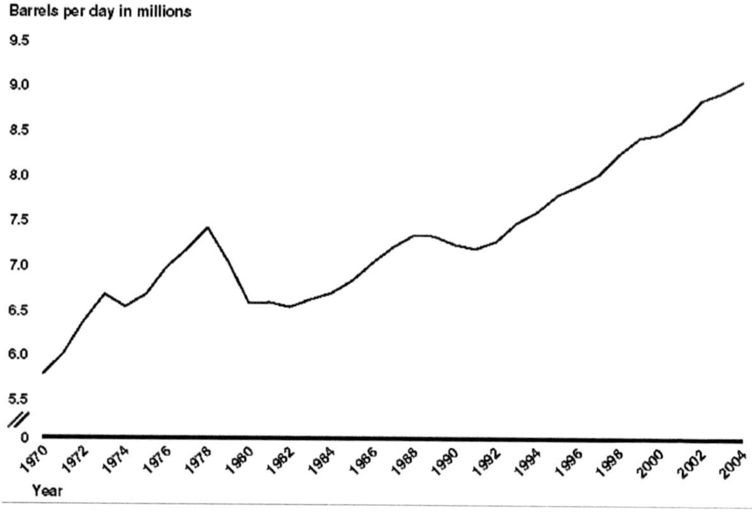

Figure 16: U.S. Gasoline Consumption (1970-2004):

Source: GAO analysis of EIA data.

Notes: One barrel of gasoline equals 42 U.S. gallons.

The United States is by far the largest consumer of gasoline in the world. As of 2001, the most recent year for which EIA data were available for international comparisons, the United States accounted for 43 percent of the world's total gasoline consumption. (See Fig. 17.)

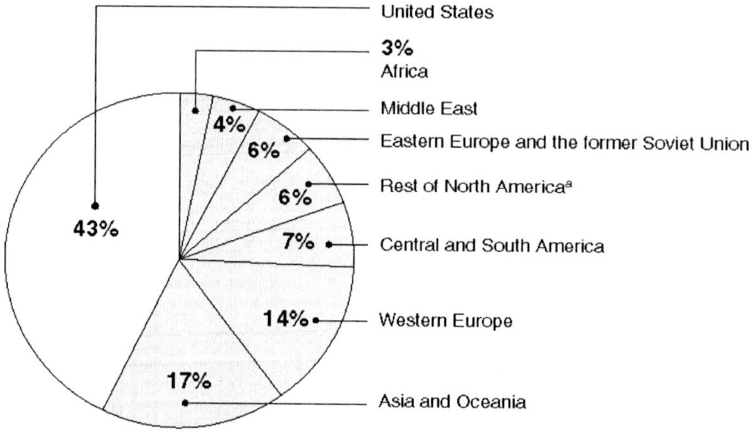

Figure 17: World Gasoline Consumption (2001):

Source: GAO analysis of EIA data.

Notes: 2001 data are the latest available.

a According to EIA, this group consists of Canada, Mexico, Bermuda, Greenland, Saint Pierre, and Miquelon, with consumption by Canada and Mexico making up the majority.

Most gasoline consumed in the United States is used for automobile transportation, particularly passenger cars, vans, pickups, and sport utility vehicles (SUV). Average fuel efficiency for passenger cars, as measured in miles per gallon of gasoline consumed, increased from about 13.5 miles per gallon in 1973 (the earliest year for which data were available) to about 22 miles per gallon in 2003. Average fuel efficiency for vans, pickups, and SUVs also increased during that period, from about 10.5 miles per gallon to 18 miles per gallon. However, since 1992, fuel efficiency performance for both vehicle categories has remained almost flat. (See Fig. 18.)

Figure 18: Fuel Efficiency of Automobiles in the United States (1973- 2003):

Source: GAO analysis of EIA data.

As Figure 19 shows, gasoline consumption by passenger cars has been relatively stable since 1980 compared with consumption by vans, pickups and SUVs, which has increased more steeply as these larger, less fuel-efficient vehicles have become a growing part of the automotive fleet. Taken together, Figures 16, 18, and 19 suggest that most of the increased U.S. gasoline consumption over the last 2 decades has been due to these larger vehicles.

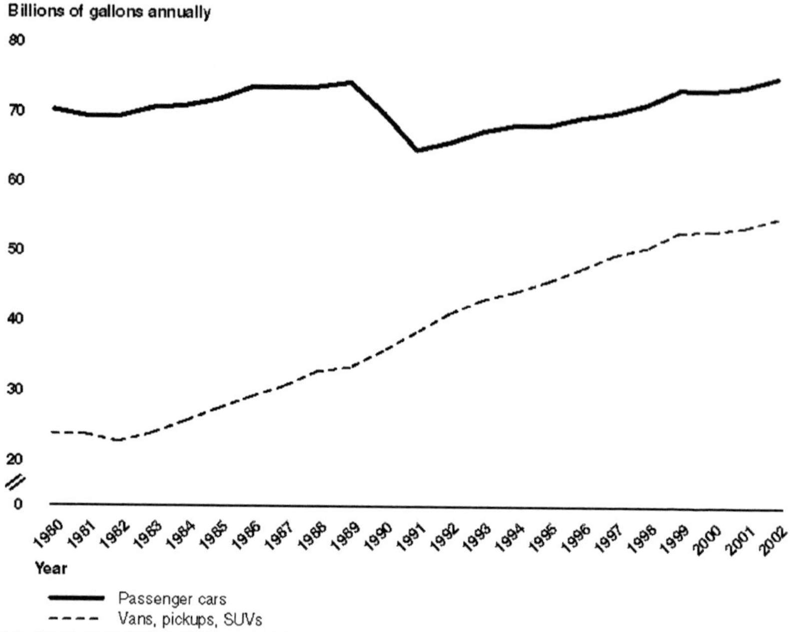

Figure 19: U.S. Gasoline Consumption by Type of Vehicle (1980-2002):
Source: GAO analysis of Department of Commerce data.

Q. Can U.S. refineries produce as much gasoline as consumers demand?

A. The number of refineries in the United States has significantly decreased since 1981 as many, mostly smaller and inefficient refineries have been shut down or mothballed. Refiners have been able to expand the capacity of existing refineries (see Fig. 20), but such "capacity creep" has not kept pace with growing demand for petroleum products.

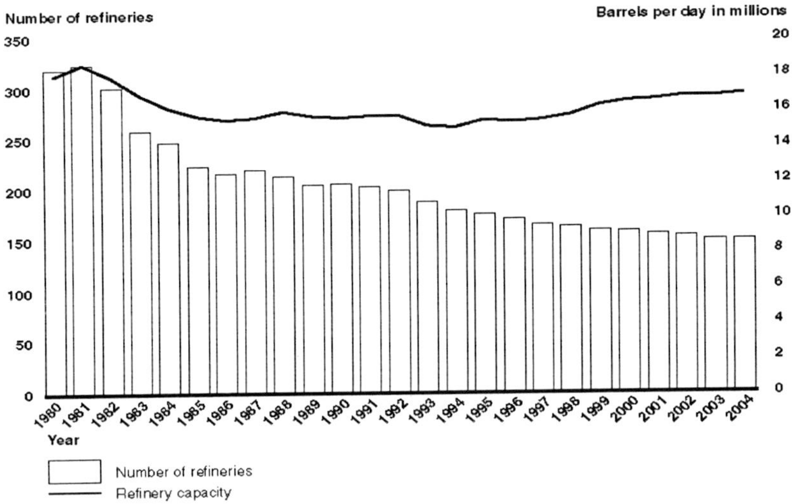

Figure 20: Refinery Capacity and Number of Refineries (1980-2004):
Source: GAO analysis of EIA data.

To meet the growing demand, U.S. refineries have been running close to their production capacity since the 1990s, as shown by the tightening balance between the amount of refinery input--such as crude oil and condensates--and refinery capacity (see Fig. 21).

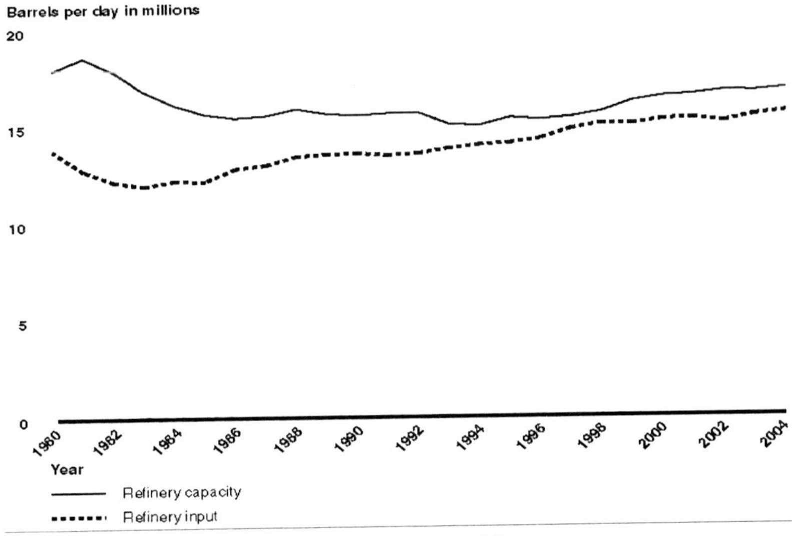

Figure 21: U.S. Refinery Capacity and Input (1980-2004):
Source: GAO analysis of EIA data.

Put differently, refineries are running at high rates of utilization. For example, U.S. refineries, on average, have been utilizing over 93 percent of their production capacity since the mid-1990s, compared with about 77.6 percent in the 1980s. When refineries are already running at such high capacity utilization rates, they have little ability to further increase production as demand increases. Also, high utilization rates can increase operating costs and lead to higher prices. We have reported that refinery capacity utilization rates of over 90 percent contributed to higher gasoline prices at the wholesale level in the second half of the 1990s.[Footnote 12] Complicating the tightening refinery capacity relative to demand is the fact that no new refineries have been built in the United States since the 1970s. This fact was noted as a continuing concern by many industry officials with whom we spoke.

Many offered explanations for the lack of new refineries, including that building a new refinery requires a capital investment of billions of dollars and that relatively low and volatile refining margins have decreased the incentive to invest in the refining sector.[Footnote 13] We did not assess whether U.S. refining margins are low. However, as shown in Figure 22, refining margins in the U.S. have been volatile and vary among regions, according to data from 1995 through 2004. According to EIA officials, differences in refining margins among regions may be explained by factors such as differences in the level of investments in refineries in those regions. For example, they noted that the margins on the West Coast are higher than in the other regions, as shown in Figure 22, perhaps reflecting the large investments that West Coast refiners made to upgrade their refineries to produce California's unique gasoline blend.

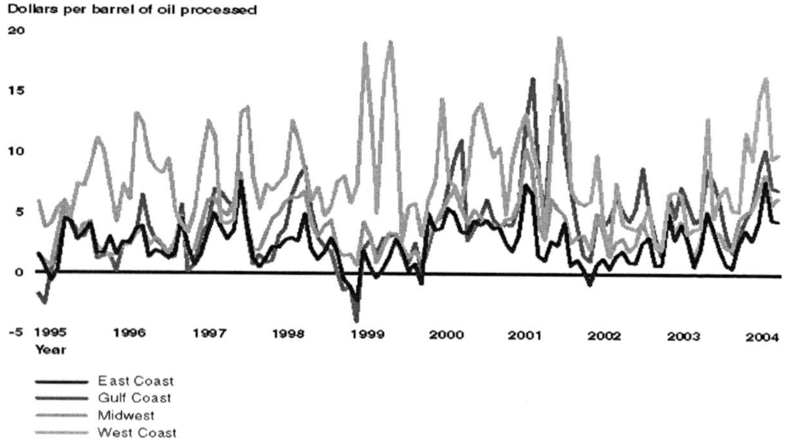

Figure 22: Refining Margins Adjusted for Inflation (1995-2004):

Source: GAO analysis of data from Muse, Stancil and Co.

Note: We used monthly refining margins adjusted to 2003 dollars using BLS's Producer Price Index for Crude Energy Materials. According to Muse, Stancil and Co., a private energy consulting firm, refining margins are an indicator of refinery earnings before interest, income taxes, depreciation and amortization.

Some industry representatives also stated that investment in refineries is discouraged by the difficulty in obtaining a permit from the relevant state or local authorities to retrofit or build a refinery in many parts of the country, owing to regulatory hurdles and local public opposition to siting refineries.

Some policymakers have offered other views, suggesting that the tight refining capacity and higher gasoline prices are not entirely due to the factors discussed above. For example, a 2002 report by the majority sta. of the Permanent Subcommittee on Investigations of the U.S. Senate stated that a lack of competition in the petroleum industry allows industry participants to make business decisions that intentionally limit refining capacity and, hence, limit gasoline supply to create opportunities for higher prices.[Footnote 14]

Evaluating the views of industry and others is beyond the scope of this primer.

Q. How does the United States balance gasoline supply and demand?

A. U.S. refiners typically achieve the balance of gasoline supply and demand not met by refinery production by drawing from gasoline in inventory and by importing gasoline. Gasoline inventory is particularly crucial to this balance because it can provide a cushion against price spikes if a refinery outage temporarily disrupts production. However, maintaining such an inventory entails costs. As Figure 23 shows, U.S. gasoline inventory has decreased substantially during the last 2-1/2 decades. While U.S. oil companies held enough gasoline in inventory to cover about 40 days of average U.S. consumption in the early 1980s, by 2004 the gasoline they held in inventory could cover less than 23 days' consumption.[Footnote 15] In part, this decline reflects a trend in business to more closely balance production with demand to reduce the cost of holding large inventories. Nonetheless, some industry watchers consider low inventories a key factor contributing to increased gasoline price volatility in recent years.[Footnote 16]

Figure 23: Days of Average U.S. Gasoline Consumption That Could Be Covered by Gasoline in Inventory (1970-2004):

Source: GAO analysis of EIA data.

Also, as demand has grown steadily and refinery capacity tightened, the United States has imported larger and larger volumes of gasoline-- especially since the mid-1990s (see Fig. 24).

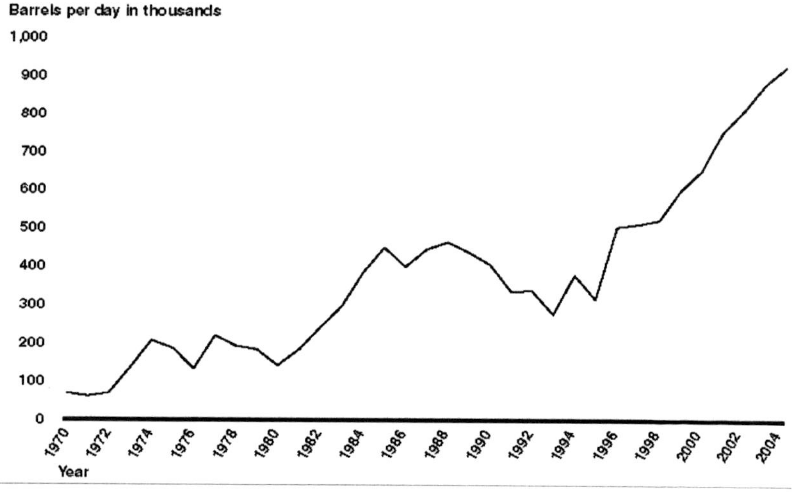

Figure 24: U.S. Gasoline Imports (1970-2004):

Source: GAO analysis of EIA data.

Note: One barrel of gasoline equals 42 U.S. gallons.

Imports accounted for about 1 percent of total U.S. gasoline consumption in 1970, rising to about 10 percent in 2004, or about 38 million gallons per day. Major sources of U.S. gasoline imports include Canada, Europe, and the Virgin Islands. According to EIA officials, this increase in imports reflects, in part, the availability of gasoline from foreign sources at lower cost than building and operating additional refinery capacity in the United States would entail, potentially dampening the overall price of gasoline for U.S. consumers. Nonetheless, because it can take several weeks for gasoline to get to the United States from some of these areas, imported gasoline may not be effective in cushioning price spikes caused by temporary supply disruptions.

As the need for imported gasoline increases, the United States may face problems in accommodating enough imports, at least in some parts of the country. According to industry representatives, marine terminals where tankers bring petroleum products, especially in California, are not equipped to handle a growing supply of gasoline from abroad.

Q. How have government requirements for special blends of gasoline affected the price of gasoline?

A. Federal and state government requirements to use special gasoline blends, often as part of overall efforts to improve air quality, can increase the costs of producing gasoline, which may be passed on to consumers in the form of higher gasoline prices. Under the Clean Air Act, as amended, the federal government and some states now require the use of a variety of gasolines, generally in areas not in compliance with federal air quality standards. The number of such special gasoline blends has grown over time and, as of the summer of 2004, there were at least 11 different blends, in addition to the more commonly used conventional gasoline. Seasonal and other factors, such as multiple octane grades, can raise this number to nearly 50 gasoline blends over the course of a year. To produce some of these gasoline blends, refiners have incurred extra costs, such as investing in additional refinery units. They also incur additional costs to produce or purchase the blending components used to make special gasoline blends. For example, the Environmental Protection Agency and the Department of Energy have estimated the cost of producing reformulated gasoline--a special-blend gasoline mandated by the federal government--to be approximately 4 to 8 cents per gallon greater than conventional gasoline. (See section 4 of this primer for further discussion of the use of special gasoline blends.) A detailed analysis of the effects of special blends of gasoline will be presented in a GAO study to be issued this year.

Q. Why are gasoline prices generally higher during certain times of the year?

A. As Figure 25 illustrates, average U.S. retail gasoline prices generally rise sharply in the spring, peaking around Memorial Day. Prices then begin to fall and stabilize somewhat in the summer-- although they remain substantially higher than in winter. Prices rise again leading up to Labor Day, after which they usually fall steeply for winter.

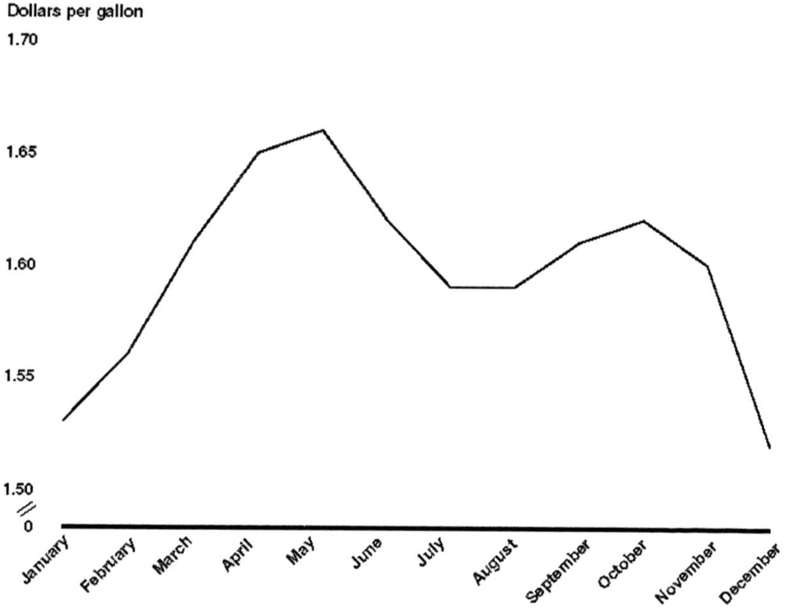

Figure 25: Average U.S. Monthly Gasoline Prices (2000-2004):

Source: GAO analysis of BLS data.

Note: Data are monthly prices for regular motor gasoline, U.S. city averages. Gasoline prices were adjusted to 2003 dollars using BLS's Consumer Price Index for All Urban Consumers: Energy.

Several factors may help explain this pattern. First, U.S. gasoline consumption fluctuates depending on the seasons. As Figure 26 illustrates using data for 2000 through 2004, within any given year, gasoline consumption in the United States is generally higher from May through early September--the summer driving season--and tends to flatten out after Labor Day. The increased summer demand puts more pressure on the supply system, and prices rise in anticipation of this pressure.

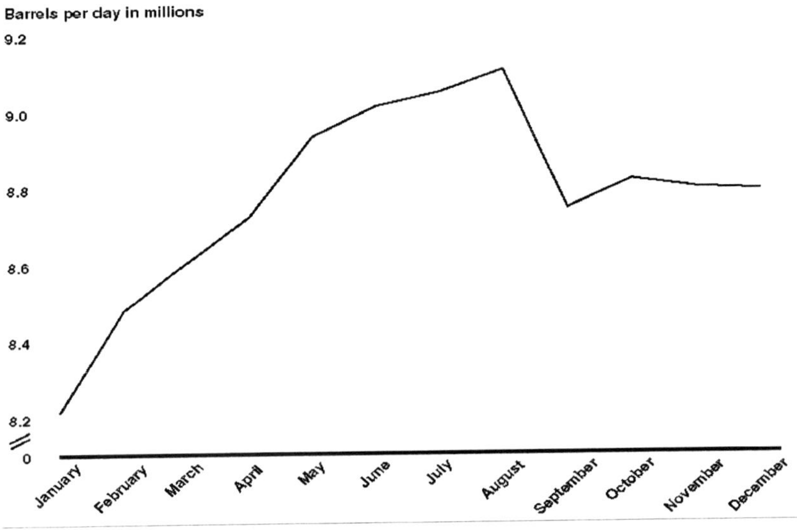

Figure 26: Average U.S. Monthly Gasoline Consumption (2000-2004):

Source: GAO analysis of EIA data.

Note: One barrel of gasoline equals 42 U.S. gallons.

The second factor is the routine maintenance work on refineries that is generally performed during the first and fourth quarters of the year. This maintenance work requires shutting down parts of refineries over a period of weeks. Although the maintenance schedule is typically planned in advance to reduce the likelihood of a supply shortfall, sometimes maintenance can take longer than expected because of unforeseen problems. Such extended outages can further reduce inventories and threaten the demand-supply balance. According to one industry representative, even a rumor about such an extended outage can trigger higher prices in the affected markets because of fears of a potential supply shortage.

The third factor is the transition from winter-grade gasoline to summer-grade gasoline. As part of efforts to improve air quality, some areas of the country are required to use a type of gasoline in the summer that has lower evaporation rates than winter-grade gasoline. To prepare for the switch from winter-grade to summer-grade gasoline, industry representatives told us that they start shrinking the inventory of winter-grade gasoline in the spring. The 2002 report by the Senate Permanent Subcommittee on Investigations summarized how the Environmental Protection Agency described the effects of low spring inventories on price as follows:

Although gasoline prices generally rise around Memorial Day, the start of the summer driving season, for the past two years spikes have occurred in various parts of the United States. These price spikes occur when gasoline inventories have become unusually low. Low gasoline inventories have occurred for a variety of reasons, including a recent trend in the petroleum industry towards reducing inventories to near the minimum operating levels. This has been particularly the case recently during the winter to summer transition. Additionally, because it costs refiners more to make summer grade fuel than winter grade fuel, competitive economic pressures lead refiners to delay this expense as long as possible.[Footnote 17]

Q. How have gasoline taxes contributed to changes in gasoline price over time?

A. Taxes are a significant component of the price of gasoline, and the average federal and state taxes on a gallon of gasoline have increased from 24.5 cents in 1990 to 37.5 cents in 2004. However, most of the increase occurred in 1994 or earlier; therefore, changes in gasoline taxes have not been a major reason for recent fluctuations in gasoline prices. As Figure 27 shows, federal gasoline taxes changed in December 1990 and again in October 1993 and have remained relatively constant at 18.4 cents per gallon since then. State taxes on gasoline vary among states (see sec. 4), but the average for all states has fluctuated slightly over time and actually decreased in 2001 from 20.17 cents per gallon to 19.07 cents per gallon. From 2001 to 2004, total federal and state taxes remained steady at about 37.5 cents per gallon,[Footnote 18] while average retail gasoline prices, adjusted for inflation, increased by about 25 percent during that time.

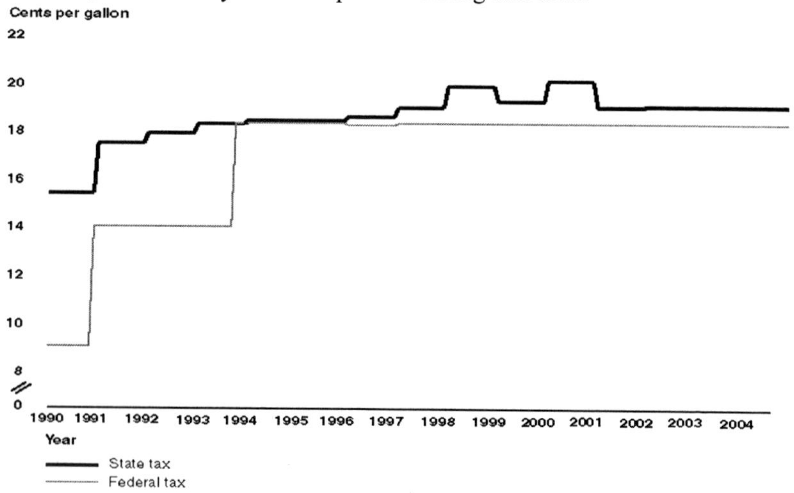

Figure 27: Federal and State Gasoline Taxes on a Gallon of Gasoline (1990-2004):

Source: GAO analysis of Federal Highway Administration data provided by EIA.

Q. How have the mergers that occurred in the petroleum industry in the 1990s affected competition and the price of gasoline?

A. Mergers between companies in the U.S. petroleum industry can reduce competition in the gasoline market. Reduced competition can lead to higher prices because, holding other things constant, the less competition there is among sellers, the more likely it is that companies can exert control over prices and raise the prices that consumers will pay. However, mergers can also lead to lower prices if industry achieves cost savings that are passed on to consumers. The Federal Trade Commission (FTC) reviews proposed mergers in the petroleum industry and has taken actions on some proposed mergers to preserve competition. For example, according to the FTC, from 1981 to 2004, 15 proposed petroleum mergers would have resulted in significant reductions in competition and harmed consumers in one or more relevant markets. Four of the mergers were abandoned or blocked as a result of FTC or court action, while in the other 11 cases, the FTC required the merging companies to divest substantial assets in the markets where competitive harm was likely to occur.

We reported in May 2004 that over 2,600 mergers have occurred in the petroleum industry since the 1990s,[Footnote 19] some involving oil companies that had previously competed with each other. (See Fig. 28.)

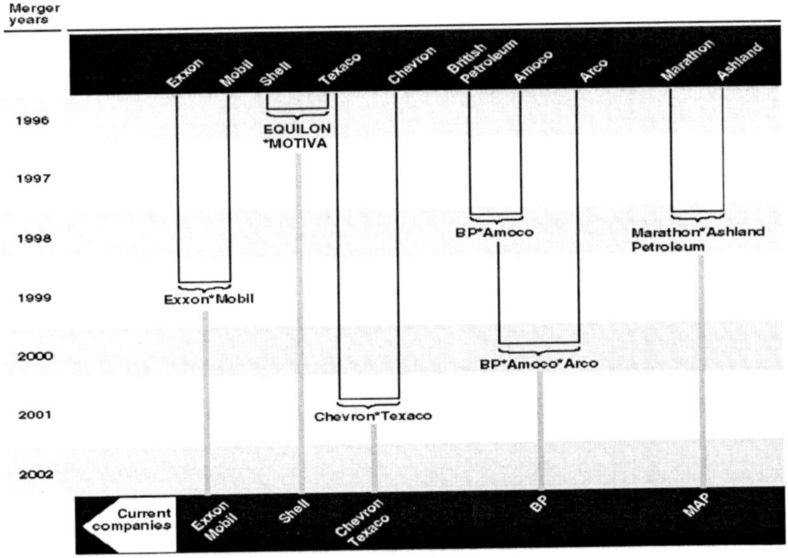

Figure 28: Selected Major Petroleum Mergers (1996-2002):

Source: GAO.

Our study found that mergers contributed to reduced competition in the industry. Moreover, the mergers we reviewed generally led to average price increases of about 1 cent to 2 cents per gallon for wholesale gasoline during the second half of the 1990s, although two of the mergers generally led to price decreases of about 1 cent per gallon, on average. Other factors discussed in this primer also contributed to price changes during the period of the study. While our study did not examine the mergers' effect on retail gasoline prices, it did note that retail gasoline prices generally follow the same trends as wholesale gasoline markets, with some lag.

SECTION 4: GEOGRAPHIC VARIATIONS IN GASOLINE PRICE:

Q. Why do gasoline prices vary from place to place?

A. Retail gasoline prices can vary from one region of the United States to another, between and within states and cities, and even within neighborhoods. One reason for these variations is proximity to the supply used to meet regional demand. Additional reasons include differences in the gasoline used--that is, conventional or special blend gasoline mandated by federal and/or state regulation to meet clean air requirements in some areas; differences in taxes at the state and/or local levels; and differences in the extent of competition in local retail markets, including competition between retailers that sell name-brand gasoline sold under the refiner's trademark such as Exxon or Shell and those selling unbranded (generic) gasoline. These reasons are discussed in further detail in the answers to subsequent questions in this section.

Q. How does proximity to the source of supply affect the price of gasoline?

A. As Figure 29 shows, most regions of the country rely on gasoline shipments from other regions to meet their gasoline demand. In September 2004, for example, only the Gulf Coast region produced significantly more gasoline than it consumed, and it was the only region to send gasoline to each of the other four regions. The Gulf Coast region's excess gasoline production capacity contributes to its generally low average gasoline prices compared with other regions of the country. For example, as shown in Figure 30, during the week of March 21, 2005, consumers in the Gulf Coast region paid an average of $2.02 per gallon for regular gasoline--the lowest price in the nation. In contrast, consumers on the West Coast paid an average of $2.25 per gallon,[Footnote 20] Midwest consumers paid an average of $2.11, and consumers on the East Coast and in the Rocky Mountains paid an average of $2.07.

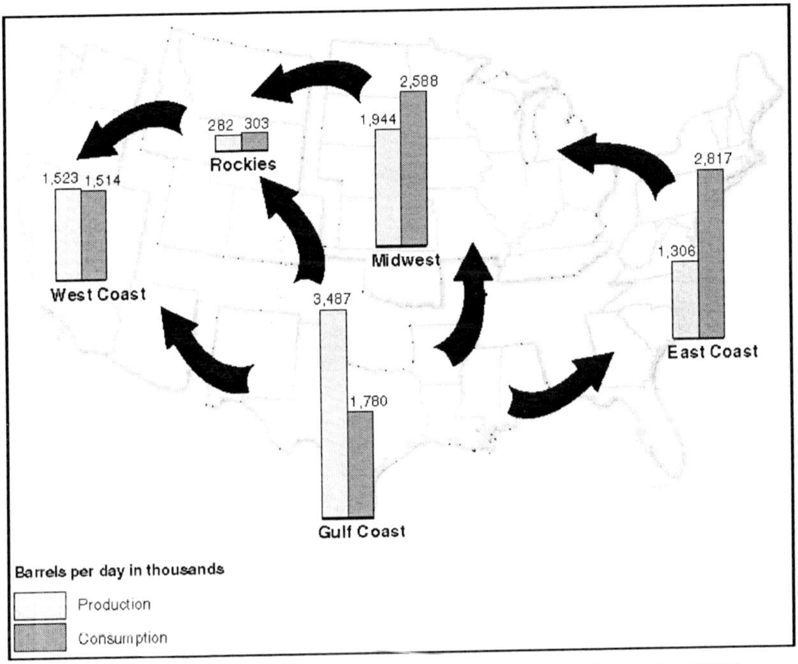

Figure 29: Gasoline Production and Consumption by U.S. Region (September 2004):

Source: GAO analysis of EIA data.

Note: Arrows represent net flow of finished motor gasoline between regions.

Q. How do differences in the types of gasoline used cause geographic variations in gasoline price?

A. Some areas of the country require the use of special blends of gasoline to meet clean air standards and provide health benefits. There were 12 distinct gasoline blends in use in various locations throughout the United States during the summer of 2004: 11 special gasoline blends and the conventional gasoline used everywhere that a special blend is not (see Fig. 31). Supplying special gasoline blends generally involves increased production, transportation, and storage costs, and these costs may be passed on to consumers as higher gasoline prices. For example, California's high gasoline price--the highest in the country- -has been partly attributed to its unique blend of gasoline. Areas that use special-blend gasoline are also more susceptible to price spikes because such requirements have made gasoline less fungible across geographic markets. In other words, because of the requirements for special gasoline blends in certain markets, refiners have limited flexibility to ship gasoline from a surplus market to a market that may be experiencing a deficit and hence price spikes due to a supply disruption. Environmental Protection Agency officials told us that while the use of special gasoline blends may have some added costs, their studies have shown that, overall, the environmental and health benefits outweigh the costs.

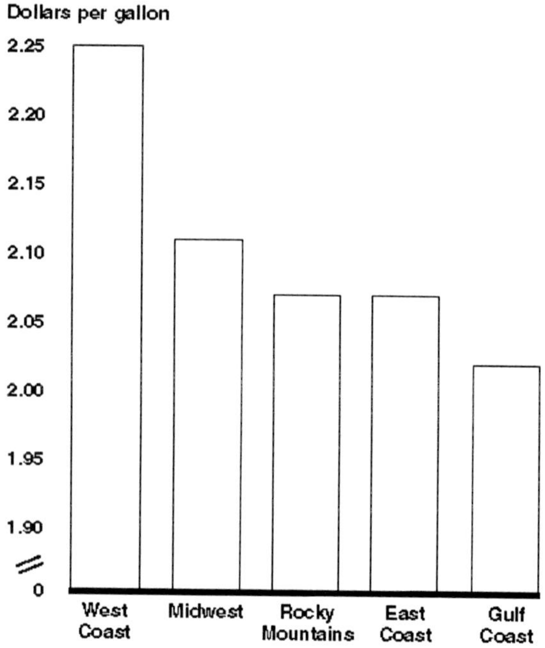

Figure 30: Average Regional Gasoline Prices (Week of March 21, 2005):

Source: GAO analysis of EIA data.

Note: Price is for regular gasoline, all formulations.

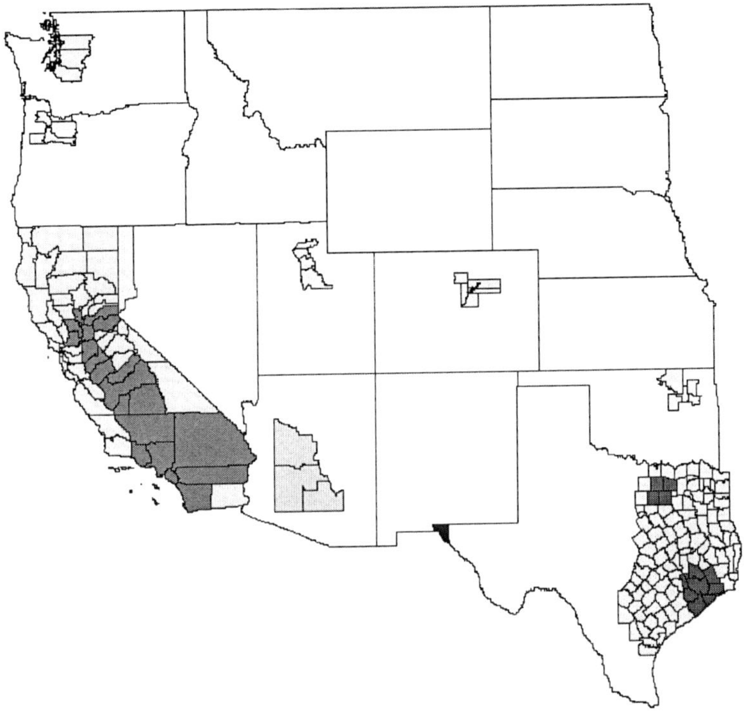

Legend: Numbers in parentheses indicate the number of counties in which the special blend was used.

7.0 RVP	(8)	7.8 RVP	(184)
7.0 RVP, 30 ppm sulfur	(44)	AZ CBG	(3)
7.2 RVP	(3)	CA CBG	(38)

Figure 31: Special Gasoline Blends in Use during the Summer of 2004:

Source: GAO analysis of data provided by EPA, ExxonMobil, the Oil Price Information Service, and state environmental agencies.

Note: Unshaded areas are where conventional gasoline is used. RVP = Reid Vapor Pressure, a measure of gasoline's tendency to evaporate. CBG = Cleaner-burning gasoline. RFG = Reformulated gasoline.

Q. How do geographic differences in taxes affect gasoline prices?

A. Differences in gasoline taxes help explain why gasoline prices vary from place to place in the United States. In addition to federal taxes, which apply across the board, states and, in some cases, local jurisdictions also impose taxes and other fees on gasoline that add to the price. Figure 32 shows total gasoline taxes for each of the 50 states and the District of Columbia, as of November 2004. New York, Hawaii, and California have the highest total gasoline taxes, while Alaska, Wyoming, and New Jersey have the lowest. While differences in taxes affect the price of gasoline, there is no consistent relationship between high taxes and high prices. For example, on March 7, 2005, gasoline cost $1.91 per gallon in North Carolina and $1.98 per gallon in Alaska, even though the taxes paid in North Carolina were almost 17 cents per gallon higher.

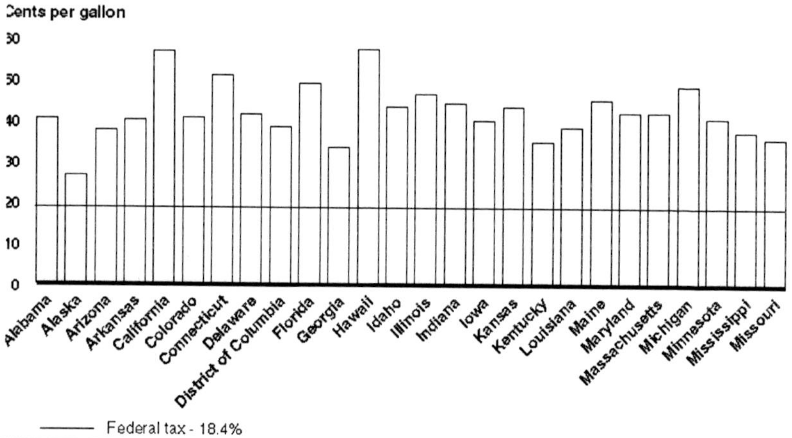

Figure 32: Estimated Total Gasoline Taxes by State (November 2004):

Source: GAO analysis of API data.

Note: According to API, these tax data include applicable state sales taxes, gross receipts taxes, and other applicable fees but largely exclude local taxes, which may average about 2 cents per gallon nationwide.

Differences in the level of taxes imposed on gasoline also help explain why U.S. consumers, on average, pay far less for gasoline than consumers in some other nations. As Figure 33 shows, the United States has, by far, the lowest average gasoline tax of the major industrialized nations, including oil-producing countries such as Canada and the United Kingdom. For these industrialized U.S. counterparts, gasoline taxes ranged from the equivalent of $1 per gallon in Canada to over $4 in the United Kingdom.

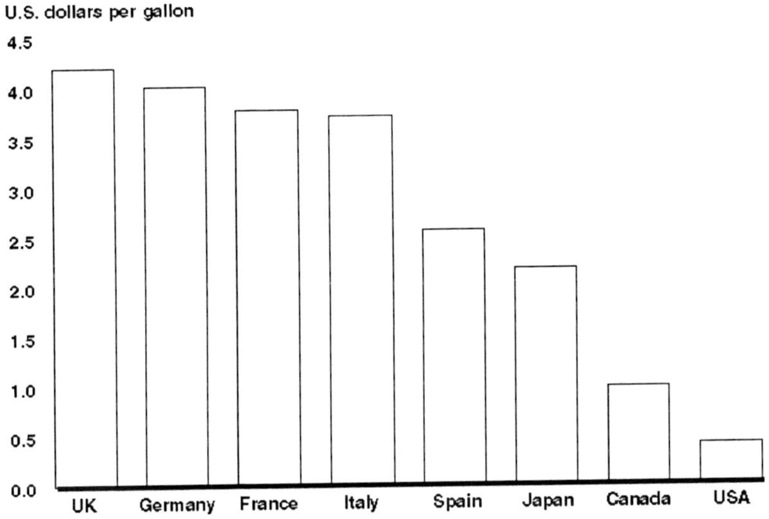

Figure 33: Gasoline Taxes in Selected Industrialized Countries (November 2004):

Source: GAO analysis of International Energy Agency data.

As Figure 34 shows, when taxes are excluded, gasoline prices in the United States are generally comparable to the prices in most of the other nations.

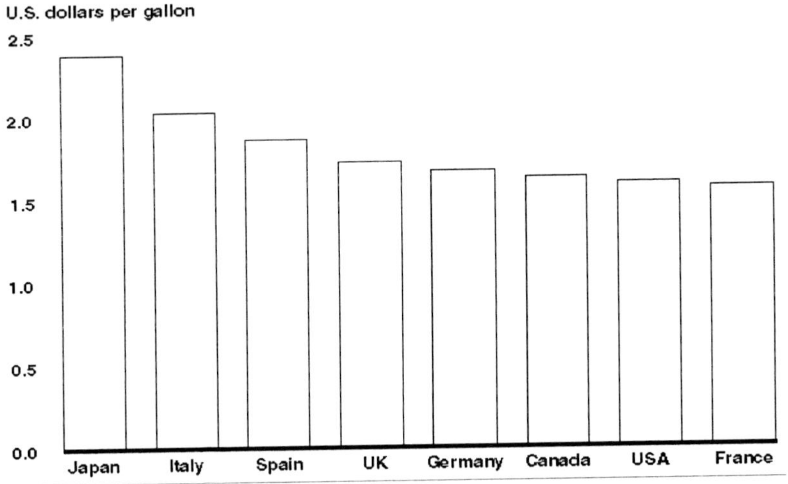

Figure 34: Average Gasoline Prices without Taxes in Selected Industrialized Countries (November 2004):

Source: GAO analysis of International Energy Agency data.

Q. How does local competition affect gasoline prices?

A. While retailers will normally seek to set their prices to cover their costs and make a profit, retail gasoline prices are strongly influenced in the short run by the extent of competition within a local market. Nationwide, the number of gasoline stations has shrunk significantly--from over 202,000 in 1994 to about 167,000 in 2003, a 17 percent decrease. These nationwide data provide some broad insight but cannot be used as a measure of retail market competition because retail gasoline markets are more local than national. Some markets have more competing gasoline stations within the vicinity than others, allowing consumers more options to shop around for lower prices. In general, areas with many competing gasoline stations tend to have lower average prices than areas with fewer competing stations, presumably because gasoline retailers closely watch the pricing by other neighboring retailers and keep the retail price competitive so as to preserve market share and profitability. For example, we previously reported that one reason why San Francisco consumers paid higher gasoline prices than those in Los Angeles, based on 1996 data, was that there were about 19 gasoline stations in San Francisco for every 100,000 people, while there were about 25 stations per 100,000 people in Los Angeles.[Footnote 21] Our study found that the high costs of land and some zoning regulations limited the siting and operation of gasoline stations in San Francisco more than in Los Angeles. Our study also found that higher per capita incomes in San Francisco, compared with Los Angeles, may have been a factor in explaining why gasoline prices were higher in San Francisco because higher incomes may make the demand for gasoline less price-sensitive.

The extent of local competition is at the heart of a practice in gasoline marketing known as "zone pricing" that helps explain why prices may differ at stations located in neighboring markets, such as neighboring towns, or between pockets of stations within the same area. As described in the 2002 Senate Permanent Subcommittee on Investigations report, oil companies typically group the retail stations that sell gasoline under the companies' trademark (brand) into geographic or market zones, which can comprise several stations or as few as one station. [Footnote 22] The companies then charge different wholesale prices to retailers by zone. Oil companies create zones essentially by determining how high a station can raise its price before it starts to lose market share. [Footnote 23] Since retailers typically seek to set their prices to cover their costs and make a profit, the wholesale price charged within a given zone plays a large part in determining the retail price charged to consumers.

Q. How does competition between name-brand and unbranded gasoline affect price?

A. Among the competitors for consumers' gasoline dollars are retailers that sell name-brand gasoline and retailers that sell unbranded gasoline. Which one you choose can affect the price you pay, and areas with more unbranded retailers generally tend to have lower prices, all things being equal.

Name-brand gasoline is generally higher priced than unbranded gasoline because its price tends to include a premium reflecting the recognized brand name, advertising costs, and guaranteed supply, among other things. Retailers of unbranded gasoline generally offer lower prices because that is how they attract customers. Unbranded retailers may also pay lower wholesale prices for their gasoline because, traditionally, they have been able to shop around in the wholesale marketplace, without any binding contracts, for the best price. However, this situation may be changing because some wholesale suppliers of unbranded gasoline are now requiring buyers to sign a binding contract to guarantee their supply. Unbranded retailers without such contracts may end up paying higher wholesale prices than name- brand retailers during periods of overall supply scarcity relative to demand, such as when there is a supply disruption caused by a pipeline or refinery breakdown.

The growing role of hypermarkets--a relatively new breed of gasoline retailers that includes such large retail warehouses as Wal-Mart and Costco--has added to the dynamics of gasoline retail competition. These hypermarkets, which sell mostly unbranded gasoline, have been gaining market share through aggressive pricing strategies. In particular, many gasoline distributors contend that hypermarkets use their gasoline as a "loss leader" to draw consumers in and thereby subsidizing gasoline sales with profits from store sales.

FOOTNOTES:

[1] This primer does not explore whether changes in gasoline prices are related to any industry anticompetitive behavior that may be prohibited by law. Such investigations are outside the scope of our study and fall under the jurisdiction of regulatory or law enforcement agencies such as the Federal Trade Commission and the Justice Department.

[2] EIA is an independent agency within the U.S. Department of Energy that collects energy data.

[3] In the industry, low-sulfur light crude oil is referred to as "light sweet," while high-sulfur heavy crude oil is referred to as "heavy sour.":

[4] Crude oil extraction is preceded by activities that include acquiring or leasing land and exploring for crude oil, but for the purposes of this study we have omitted these activities.

[5] EIA uses tax data from the American Petroleum Institute (API) for its tax analysis. According to API, these data include applicable state sales taxes, gross receipts taxes, and other applicable fees but largely exclude local taxes, which may average about 2 cents per gallon nationwide.

[6] The National Commission on Energy Policy is a bipartisan commission of industry, academic, and government officials.

[7] As of April 2005, OPEC's members are Algeria, Indonesia, Iran, Iraq, Kuwait, Libya, Nigeria, Qatar, Saudi Arabia, the United Arab Emirates, and Venezuela.

[8] Persian Gulf countries include Bahrain, Iran, Iraq, Kuwait, Qatar, Saudi Arabia, and United Arab Emirates. According to EIA data, the United States did not import crude oil or petroleum products from Bahrain or Iran.

[9] OPEC's "basket" price, which was introduced in 1987, is a weighted average price of seven crude oils from Algeria, Dubai, Indonesia, Mexico, Nigeria, Saudi Arabia, and Venezuela.

[10] Because oil is traded in U.S. dollars worldwide, the current fall in the dollar value relative to other major currencies may be affecting the overall purchasing power of OPEC members, who receive their oil proceeds in dollars, and it is not clear if and how that might be influencing their oil pricing policy.

[11] The other futures markets where oil is traded are the International Petroleum Exchange of London and the Singapore Mercantile Exchange in Singapore.

[12] See GAO, Energy Markets: Effects of Mergers and Market Concentration in the U.S. Petroleum Industry, GAO-04-96 (Washington, D.C.: May 17, 2004).

[13] A recent study by the National Petroleum Council reported that from 1981 through 2002, returns in the petroleum industry as a whole averaged 11.3 percent, slightly lower than the 12.2 percent average for the companies in the Standard and Poor's 500 index. The study also indicated that returns in the refining and marketing sector had historically been lower than in the rest of the petroleum industry. See National Petroleum Council, Observations on Petroleum Product Supply a Supplement to the NPC Reports: U.S. Petroleum Product Supply--Inventory Dynamics, 1998 and U.S. Petroleum Refining--Assuring the Adequacy and Affordability of Cleaner Fuels, 2000 (Washington, D.C., December 2004), 1-14.

[14] United States Senate Permanent Subcommittee on Investigations, Committee on Governmental Affairs, Gas Prices: How Are They Really

Set? (released in conjunction with the Permanent Subcommittee on Investigations hearings on April 30 and May 2, 2002).

[15] Moreover, as noted by EIA officials, not all gasoline in inventory may be available for use because a minimum level must be left in the storage and transportation system, such as in pipelines, at all times to keep the system operating.

[16] See, for example, these three citations from the United States Senate Permanent Subcommittee on Investigations, Committee on Governmental Affairs, Gas Prices: How Are They Really Set? (1) "As EIA has pointed out on numerous occasions, very low gasoline stocks, combined with a market short on crude oil, generates an environment ripe for price volatility, both during the spring and peak summer periods." EIA, testimony before the Committee on Energy and Commerce (May 15, 2001), 54, fn. 96. (2) "Stocks are the key factor in policy responses to market power where supply is inelastic. Every investigation of every product spike in the past several years points to unusually low stock as a primary driver of price shocks." M. Cooper, Consumer Federation of America, Ending the Gasoline Price Spiral (July 2001), 10-11. (3) "The key determinant of the observed price is the amount of inventories held by processors and consumers." P.K. Verleger Jr., World Oil Markets: Changing Structure and Greater Price Volatility Causing the Third Petro-Recession, draft (April 2001).

[17] Gas Prices: How Are They Really Set?

[18] The state tax data used here do not include applicable sales taxes and other fees, which are included in API's November 2004 estimates of taxes, presented in Figure 7. API does not currently have such data.

[19] GAO-04-96.

[20] Gasoline prices in California are among the highest in the nation, even though most of the gasoline consumed in the state is refined there. The high prices are due mainly to complications related to supplying California's unique blend of gasoline. For more information on the impact of California's special gasoline blend on the state's gasoline prices, see, for example, GAO, Motor Fuels: California Gasoline Price Behavior, GAO/RCED-00-121 (Washington, D.C.: Apr. 28, 2000).

[21] GAO/RCED-00-121.

[22] Gas Prices: How Are They Really Set?

[23] Gas Prices: How Are They Really Set?

INDEX

M

N

O

P

Q

R

S